T0139855

Lecture Notes on Data Engineering and Communications Technologies

Volume 28

Series editor

Fatos Xhafa, Technical University of Catalonia, Barcelona, Spain
e-mail: fatos@cs.upc.edu

The aim of the book series is to present cutting edge engineering approaches to data technologies and communications. It will publish latest advances on the engineering task of building and deploying distributed, scalable and reliable data infrastructures and communication systems.

The series will have a prominent applied focus on data technologies and communications with aim to promote the bridging from fundamental research on data science and networking to data engineering and communications that lead to industry products, business knowledge and standardisation.

** Indexing: The books of this series are submitted to ISI Proceedings, MetaPress, Springerlink and DBLP **

More information about this series at http://www.springer.com/series/15362

Nabendu Chaki · Nagaraju Devarakonda ·
Anirban Sarkar · Narayan C. Debnath
Editors

Proceedings of International Conference on Computational Intelligence and Data Engineering

Proceedings of ICCIDE 2018

 Springer

Editors
Nabendu Chaki
Department of Computer Science
and Engineering
University of Calcutta
Kolkata, West Bengal, India

Anirban Sarkar
National Institute of Technology Durgapur
Durgapur, West Bengal, India

Nagaraju Devarakonda
Department of Information Technology
Lakireddy Bali Reddy College
of Engineering
Mylavaram, Andhra Pradesh, India

Narayan C. Debnath
Eastern International University
Thu Dau Mot, Binh Duong, Vietnam

ISSN 2367-4512 ISSN 2367-4520 (electronic)
Lecture Notes on Data Engineering and Communications Technologies
ISBN 978-981-13-6458-7 ISBN 978-981-13-6459-4 (eBook)
https://doi.org/10.1007/978-981-13-6459-4

Library of Congress Control Number: 2019930971

This Springer imprint is published by the registered company Springer Nature Singapore Pte Ltd.
The registered company address is: 152 Beach Road, #21-01/04 Gateway East, Singapore 189721,
Singapore

Preface

The Second International Conference on Computational Intelligence and Data Engineering (ICCIDE 2018) took place on 28 and 29 September 2018 in Vignan's Foundation for Science, Technology & Research (Vignan University) at Vadlamudi, Guntur District, Andhra Pradesh, India. The academic partners for ICCIDE 2018 are University of Calcutta; Lakireddy Bali Reddy College of Engineering (LBRCE), Andhra Pradesh, India; Winona State University, Minnesota, USA; NIT Durgapur, West Bengal, India.

The First International Conference on Computational Intelligence and Data Engineering (ICCIDE 2017) was held at Lakireddy Bali Reddy College of Engineering on 14 and 15 July 2017, and the proceedings of the conference were published in Lecture Notes on Data Engineering and Communications Technologies (LNDECT) series of Springer.

https://www.springer.com/in/book/9789811063183.

ICCIDE is conceived as a forum for presenting and exchanging ideas, which aims at resulting in high-quality research in cutting-edge technologies and most happening areas of computational intelligence and data engineering. The conference solicited the latest research ideas on computational intelligence and data engineering, thus inviting researchers working in the domains of machine learning, Bayesian network, computational paradigms and computational complexity, intelligent architectures, intelligent applications, natural language processing, grey theory, swarm intelligence, semantic web, knowledge representation, data models, ubiquitous data management, mobile databases, data provenance, workflows, scientific data management and security.

The chapters of this book are comprised of high-quality research submissions done by scholars from India and various countries, which is achieved by the endless efforts of the programme committee members and Springer. A thorough peer-review process has been carried out by the PC members and by the external reviewers. The reviewers focused on the novelty of each research submission by keenly observing the contribution, technical content, structural organization and clarity of presentation. The entire process of paper submission, review and acceptance process was done electronically.

The technical programme committee eventually could identify 25 papers for the publication out of 212 submissions. The resulting acceptance ratio is 11.79%, which is a quite good number in Springer series of the conference.

The programme also includes keynotes by Professor Yingxu Wang from University of Calgary, Canada; Professor Newton Howard from University of Oxford, UK; Professor Fiorini Rodolfo from Politecnico di Milano University, Italy; Professor David Westwick from Schulich School of Engineering University of Calgary, Canada; Professor C. Chandra Sekhar from IIT Madras, Chennai, India; and Aninda Bose, Senior Editor, Springer, New Delhi, India.

We eventually extend our gratitude to all the members of the programme committee and the external reviewers for their excellent and time-bound review work. We thank all the sponsors, who have come forward towards organizing this symposium. We submit our indebtedness to Dr. Lavu Rathaiah, Chairman; Sri L. Sri Krishna Devarayalu, Vice Chairman, Vignan's Group of Institutions; Dr. K. Ramamurty Naidu, Chancellor, VFSTR; Dr. M. Y. S. Prasad, Vice Chancellor, VFSTR, Vadlamudi. We appreciate the efforts done by Dr. K. Krishna Kishore and Mr. U. Janardhan Reddy of VFSTR for the smooth conduct of the conference. We are thankful to the entire management of Vignan's Group of Institutions for their patronage and continual support to make the event successful.

We appreciate the initiative and support from Mr. Aninda Bose and his colleagues in Springer Nature for their strong support towards publishing this volume in the Lecture Notes on Data Engineering and Communications Technologies (LNDECT) series of Springer Nature. Finally, we thank all the authors without whom the conference would not have reached the expected standards.

Vijayawada, India   Nabendu Chaki
October 2018   Nagaraju Devarakonda
                                                          Narayan C. Debnath
                                                            Anirban Sarkar

Contents

About the Editors

Nabendu Chaki is a Professor at the Department Computer Science & Engineering, University of Calcutta, Kolkata, India. Dr. Chaki graduated in Physics from the legendary Presidency College in Kolkata and then in Computer Science & Engineering from the University of Calcutta. He completed his Ph.D. in 2000 at Jadavpur University, India. He shares 6 international patents, including 4 U.S. patents with his students. Prof. Chaki has been active in developing international standards for software engineering and cloud computing as a part of the Global Directory (GD) for ISO-IEC. As well as editing more than 25 books, Nabendu has authored 6 text- and research books and more than 150 Scopus Indexed research papers in journals and at international conferences. His areas of research include distributed systems, image processing and software engineering. Dr. Chaki has served as a member of the research faculty in the Ph.D. program in Software Engineering in the U.S. Naval Postgraduate School, Monterey, CA. He is a visiting faculty member of many Universities in India and abroad. In addition to being on the editorial board for several international journals, he has also served on the committees of over 50 international conferences. Prof. Chaki is the founder Chair of the ACM Professional Chapter in Kolkata.

Nagaraju Devarakonda received his B.Tech. from Sri Venkateswara University, M.Tech. from Jawaharlal Nehru University, New Delhi and Ph.D. from Jawaharlal Nehru Technological University, Hyderabad. He has published over 50 research papers in international conferences and journals. He is currently working as a Professor and Head of the IT Department at Lakireddy Bali Reddy College of Engineering. His research areas are data mining, soft computing, machine learning, and pattern recognition.

Anirban Sarkar completed his M.C.A. at University of Madras, Chennai, India and Ph.D. in Computer Science & Engineering from the National Institute of Technology (NIT), Durgapur, India. His areas of research include cloud computing, service-oriented computing, software engineering and data mining. He has more than 17 years of teaching experience in these areas. He has published over 100 papers in journals and conference proceedings.

Narayan C. Debnath is the Founding Dean of the School of Computing and Information Technology and the Head of the Department of Software Engineering at the Eastern International University, Vietnam. He is also serving as the Director of the International Society for Computers and Their Applications (ISCA). Dr. Debnath received doctorate degrees in Computer Science and in Applied Physics. He served as the President, Vice President, and Conference Coordinator of the International Society for Computers and Their Applications (ISCA), and has been a member of the ISCA Board of Directors since 2001. He has made original research contributions in software engineering and development, software testing and quality assurance, software models, metrics and tools, and information science, technology and management. He is the author or co-author of over 425 publications in numerous refereed journals and conference proceedings in the fields of computer science, information science, information technology, system sciences, mathematics, and electrical engineering.

Performance Evaluation of Guest Operating System Following Obliteration of Pre-installed Features

Osho Agyeya, Prateek Singh, Shivam Prasad, Supragya Raj and J. V. Thomas Abraham

Abstract Open-source OSs like Linux provide us means to remove certain features from the guest OS in order to increase its performance manifold, which leads to our own fine-tuned version of guest OS. This paper is aimed at studying the parameters affecting performance of guest OS and editing those features to produce its efficient version.

Keywords Linux · Operating system · Virtual machine · Guest OS · Host OS · Benchmarking · NMON

1 Introduction

A virtual machine allows the host OS to set up an environment in which a guest OS can be run along with several other applications. The host OS creates an illusion such that the guest OS considers the hardware resources as native.

This paper dives deep into the ways by which certain components of Ubuntu OS (running as guest OS), which if removed, alter the system performance consequently.

O. Agyeya (✉) · P. Singh · S. Prasad · S. Raj · J. V. Thomas Abraham
Department of Computer Science and Engineering, VIT University Chennai Campus,
Vandalur-Kelambakkam Road, Chennai, Tamil Nadu, India
e-mail: osho.agyeya2015@vit.ac.in; oshoagyeya123@gmail.com

P. Singh
e-mail: prateeksingh0001@gmail.com

S. Prasad
e-mail: shivam13juna@gmail.com

S. Raj
e-mail: supragyaraj@gmail.com

J. V. Thomas Abraham
e-mail: thomasabraham.jv@vit.ac.in

© Springer Nature Singapore Pte Ltd. 2019
N. Chaki et al. (eds.), *Proceedings of International Conference on Computational Intelligence and Data Engineering*, Lecture Notes on Data Engineering and Communications Technologies 28, https://doi.org/10.1007/978-981-13-6459-4_1

2 Tools Used

2.1 NMON

This tool is meant for benchmarking purposes which provides great deal of information using single command, either real time or saving data to a file [1].

2.2 VMware

This was used to create several VMs, with each VM having some feature of the guest OS removed. VMware is a non-free virtual machine application, which supports Ubuntu as both host and guest operating system [2].

2.3 Ubuntu 14.04 LTS

It was the guest OS for analysis. A default installation consists of general utilities and system tools [3]. Further details of the parameters of the VM are as follows: Memory: 7.8 GB—Processor: Intel®Core™i7-6700HQ2.60 GHz—OS type: 64 bits—Disk: 18.9 GB.

2.4 NMON Analyzer

This tool converts NMON data file into human-readable graph format stored in spreadsheets [4].

3 Methodology

NMON values for various parameters are recorded for the default OS. Then, a component is removed followed by recording values again. The graphs are observed for the new data and conclusions are drawn. The time delay and frequency of running can be specified in the terminal. NMON process needs to be killed later explicitly.

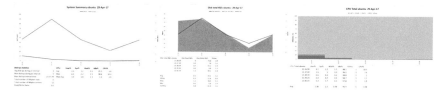

Fig. 1 NMON graphs for original system

4 Original System

The System Summary graph can be used to observe the general trends in CPU% + IO/sec(user + system) along with detailed information about disk transactions per second. The relevant parameter values can be observed from excel snap above. The disk usage (Disk total KB/s) for the system can be broken down primarily into the frequency of read and write operations. The detailed analysis of user CPU usage, system CPU usage, waiting %, idle%, and busy% can be determined from CPU Total graph. *The comparisons for the three graphs have been listed separately as 3 bullet points in each subsection* (Fig. 1).

5 Removal Stage 1: Empathy, EOG, Evince

5.1 Empathy

This offers instant messaging over audio, video, text along with file sharing [5].

5.2 EOG

Eye of GNOME serves as an image viewer for GNOME distro, equivalent to photographs app for Windows [6].

5.3 Evince

Evince is a PDF document viewer that can be used for viewing images also [7]. It is similar to Adobe Reader.

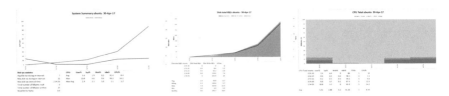

Fig. 2 Results obtained after stage 1 removal

5.4   Results from Fig. 2

- On comparing the current graph with that of original, it is observed that there is a significant decrease in the amount of IO/sec initially. The IO/sec curve is strictly increasing which reaches a peak value nearly the same as that of the original system. The CPU% curve is decreasing for the original system but in this case, the curve decreases initially to finally turn into an increasing curve with a much higher peak value of nearly five units. Thus, we can conclude that on an average, the IO operations of the system have been suppressed from increasing in the initial stages.
- The area covered by orange color at the end shows that the disk-writing activity has been delayed.
- Initially, the values for all parameters are same as before but after a time gap, the values for all the parameters drop and rise again. The peak in user CPU line at the end of the graph shows increased CPU utilization.

6   Removal Stage 2: LIBREOFFICE COMPLETE PACKAGE

LibreOffice is open-source and free office suite [8], containing tools for document, spreadsheet, and presentation creation with database analysis.

6.1   Results from Fig. 3

- The graph has two peaks for IO/sec with the maximum peak crossing the 90-unit mark. However, this graph has a drop followed by a subsequent rise, unlike the original system where only a major fall is observed. Thus, the IO operations have been delayed for an intermediate step.
- Disk write speed is an increasing curve while the original curve is falling. The area covered by the disk write KB/sec has greater value than before.

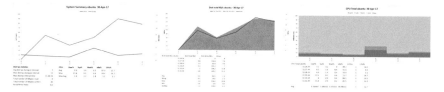

Fig. 3 Results obtained after stage 2 removal

Fig. 4 Results obtained after stage 3 removal

– This graph is different as initially, there is decreasing curve but a peak is obtained in between where maximum peak value for sys% beats the original system. User% never becomes zero, thus certain user operations keep the CPU utilized always.

7 Removal Stage 3: PRELOAD

It runs in background keeping track of most frequently used apps to cache them, thereby decreasing load time [9]. It prevents RAM from sitting idle and boosts performance.

7.1 Results from Fig. 4

– A constant fall in values is followed by a subsequent constant curve. However, at the end, there is a spike in the curve denoting sudden rise in IO operations. Similar trend is observed for CPU% too.
– There is a drop in disk usage initially. The area covered by orange is the least denoting that disk write has nearly diminished for the initial stages of the graph.
– There is a stark peak in the ending which was absent in the original graph. The peak value reaches to a value of 30 units in this case as compared to 10 units in the original case. In the intermediate stages, the CPU is underutilized, indicating a decrease in performance.

8   Removal Stage 4: GCC, LIBC-BIN, LIBGLIB2.0-0, UNITY TWEAK TOOL, AND BAMFDAEMON

8.1   *GCC*

It used as GNU C compiler platform which helps in compiling the C programs and also helps to generate an intermediate file to be executed.

8.2   *LIBGLIB 2.0-0/LIBC-BIN*

These are libraries/programs for GNU-C.

8.3   *Bamfdaemon*

It's main feature is that ATL + TAB switching of tabs. This functionality is just for extra ease.

8.4   *Unity Tweak Tool*

It is a settings manager for the Unity desktop, and it provides easy-to-use interface to access some setting of desktop.

8.5   *Results from Fig. 5*

– The graph for IO is basically the original graph that has been shifted along the x-axis. Therefore, the crests and troughs have been delayed in this case. The graph for CPU has a distant peak, which occurs at the same time.
– Areas occupied by both disk write KB/s and disk read KB/s are high and have similar configuration.
– There are three distinct peaks, each peak increasing in magnitude subsequently. The user % utilization of the CPU is dominant. The wait% is maximum in this case.

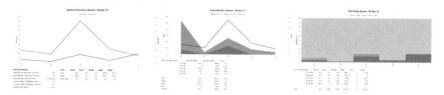

Fig. 5 Results obtained after stage 4 removal

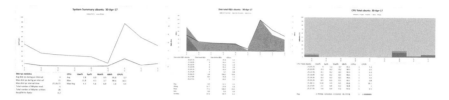

Fig. 6 Results obtained after stage 5 removal

9   Removal Stage 5: LYNX INSTEAD OF FIREFOX

It is to be used as a replacement for Firefox.

9.1   Results from Fig. 6

– A peak is observed at quite a later stage in the graph as compared to the original graph both for I/O curve and CPU curve. If the user wants to keep the load minimum in the initial stages of boot, then this component should be removed.
– In the initial stages, the disk read KB/s is significant, however, it goes down significantly later.
– There is long period of time for which there is no utilization of the CPU. The CPU is usually idle for the entire record cycle.

10   Removal Stage 6: SOUND DRIVER, DESKTOP DRIVERS, GUI DRIVERS

10.1   Results from Fig. 7

– For the first time, the curves for both IO/sec and CPU intersect thrice. The IO/sec has a distinct peak at 17.39 system time.

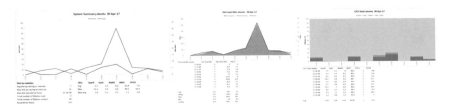

Fig. 7 Results obtained after stage 6 removal

- The curve for disk read KB/s is nearly absent. The disk write KB/s curve closely follows the IO/sec graph.
- The graph has the maximum number of distinct blocks. The troughs are characterized by 0 values for all the parameters.

11 Removal Stage 7: NTFS-3G, NETWORK-MANAGERGNOME, MYSQL-COMMON

11.1 NTFS-3G

This is an open-source version of MS Windows NTFS, with support for read/write [10].

11.2 BINARY PACKAGE "MYSQL-COMMON" IN UBUNTU

It is an agile, multiuser, and sturdy multi-threaded SQL database server.

11.3 Network Manager Gnome

It coordinates all the network devices and connections while trying to maintain active network connectivity.

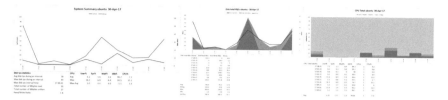

Fig. 8 Results obtained after stage 7 removal

11.4 Results from Fig. 8

- The curve of IO/sec nearly rests on the curve for CPU%. The CPU% also goes through many crests and troughs. The IO/sec acts as a strict upper bound for CPU%.
- The graph for disk read KB/sec and disk write KB/sec is nearly similar. The former is always less than or equal to the latter.
- The graph shows a segment where all parameter values are 0. Apart from the segment mentioned above, the graph is an alternative sequence of crests and troughs.

12 Conclusion

After examining the removal of all the above components, a virtual machine with maximum performance, least size can be constructed for simple general purpose performance. The components that can be removed fall under:

12.1 Components to Retain

NTFS-3G as a file system is imperative for Ubuntu. If the Internet is needed, *NETWORK MANAGER GNOME* keeps on checking the system for network connectivity. *Desktop and GUI drivers* are essential for proper interfacing with the system. *Lynx* as a web browser is smaller in size than Firefox and offers fluid CLI if only text-based surfing is needed.*GCC* library is imperative for compilation purposes. *PRELOAD* is a must since it keeps on caching the frequently used applications.

12.2   Components that May Be Removed

Empathy can be removed since instant messaging can be done online. *EOG* maybe removed since a web browser can be used for viewing local images. *Evince* is also not needed since a web browser can be used as a PDF reader also. Google Docs, spreadsheets, and Prezi are better alternatives to the entire *Libre Office Package*. It keeps running in background and checks for updates which slows the system. *Bamfdeamon* is an unnecessary add-on. *Unity tweak* just makes the user interface more accessible. Obsolete drivers like that for *floppy disk* maybe removed.

Having done the above steps, a faster, lighter, more portable, and smaller version of virtual machine is obtained that takes a lot less time to boot/restore.

References

1. nmon for Linux, http://nmon.sourceforge.net/pmwiki.php
2. VMware, https://en.wikipedia.org/wiki/VMware
3. https://www.coursehero.com/file/p4qm79t/10-it-will-move-to-the-GNOME-3-desktopinstead-as-work-on-Unity-endsUbuntu-is/
4. Griffiths N. developersWorks, I.B.M.: nmon performance: a free tool to analyze AIX and Linux performance. Copyright IBM Corporation
5. GNOME multi-protocol chat and call client, https://apps.ubuntu.com/cat/applications/raring/empathy/
6. Eye of GNOME graphics viewer program, https://apps.ubuntu.com/cat/applications/raring/eog/
7. Evince Document Viewer, https://help.ubuntu.com/community/Evince
8. LibreOffice, https://en.wikipedia.org/wiki/LibreOffice
9. Drastically Speed up your Linux System with Preload, https://www.linux.com/news/drastically-speed-your-linux-system-preload
10. NTFS-3G, https://en.wikipedia.org/wiki/NTFS-3G

Analysis of Cell Morphology, Vitality, and Motility: A Study Related to Analysis of Human Sperm

Raghavendra Maggavi, Sanjay Pujari and C. N. Vijaykumar

Abstract Attributes such as the number of sperm with motility, morphology, and vitality is necessary for successful fertilization and term pregnancy in a semen sample. Many of the vitality procedures used to stain semen result in coloring of the live/dead sperm. But they are not only costly but in addition, we need fluorescent microscope which is investment of another 10 lakh extra, moreover system fails to distinguish their individual morphology, that play a key role in the fertilization process. In the first stage, we measured sperm morphology with normal/abnormal status using a Papanicolaou stain. In the second stage, evaluated sperm vitality using Eosin–Nigrosin dye which is cheaper compared to methods used for coloring of the live/dead sperm and also measures the morphology of the individual sperm with normal/abnormal status. In the third stage, measured sperm motility along with their paths with identification numbers automatically from a video taken by research microscope. All the above methods are implemented using IMAGEJ (Image Processing using Java) open-source software.

Keywords CASA · FFMPEG · WHO · IVF · ICSI · IMSI

1 Introduction

Routine semen analysis is the most important source of information regarding the fertility status of male is more an assessment of the potential for fertility rather than a test for its fertilizing capacity. Therefore, more specialized tests are necessary to

R. Maggavi (✉)
VTU, Belagavi, India
e-mail: rrmaggavi@gmail.com

S. Pujari
Angadi Institute of Technology and Management, Belagavi, India
e-mail: sapujari@rediffmail.com

C. N. Vijaykumar
Belagavi, India
e-mail: vzchelur@yahoo.com

© Springer Nature Singapore Pte Ltd. 2019
N. Chaki et al. (eds.), *Proceedings of International Conference on Computational Intelligence and Data Engineering*, Lecture Notes on Data Engineering and Communications Technologies 28, https://doi.org/10.1007/978-981-13-6459-4_2

11

determine conclusively the fertilizing capacity of the sperms. In basic semen analysis and many research-oriented investigations, it is important to distinguish between dead and live spermatozoa. A proportion of live spermatozoa can be identified by their motility, while the ability to distinguish live, immotile spermatozoa from dead spermatozoa. Normal spermatozoa exhibit motility ranging from fast progressive to slow progressive. The spermatozoon moves at a speed of 25 μ/s or 3 mm/h. A normal ejaculate has at least 25% fast progressive or 50% (fast + slow spermatozoon) [1].

2 Materials and Methods

2.1 Sperm Head Morphology

Assessment of sperm morphology is carried out using Papanicolaou stain [1]. Papanicolaou staining procedure has been widely adopted, as this technique allows a detailed examination of the nuclear chromatin pattern. By smearing the semen sample on a glass slide, all the sperms were allowed to lie in an equal lane.

These slides were fixed and air dried. Prepared slides were analyzed under microscope with 40× objective and screened for various morphological abnormalities of sperm. The first stage of work carried out to examine morphometric parameters of the human sperm head. Major steps explained below.

- Acquisition of images;
- RGB to gray conversion, resize, calibration in μm;
- Corrects for intensity fluctuations by normalizing the images of a stack to the same mean intensity using bleach correction macro;
- Enhancement of images so that objects can be extracted from the background;
- Sperm heads will be extracted using head extraction algorithm;
- Calculation of morphometric parameters;
- Apply boundary conditions on morphometric parameters as per WHO guidelines and classify normal/abnormal form.

Normal morphometric values used in the study are computed at 75% confidence interval as normal values mentioned in WHO manual [2].

Results

Figure 1a indicates input image stack which is taken from digital microscope camera ProgRes-CT3 attached to Olympus BX-41 system microscope under 40× magnifications in the bright field mode. Figure 1b describes output image with labels, which classify the sperm with colors. Red color sperm indicates abnormal and pink color sperm indicates normal. Figure 1c points out the morphology status which is generated automatically using morphology algorithm.

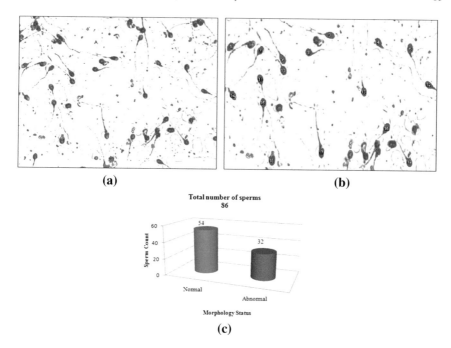

Fig. 1 **a** Input image stack **b** Output image stack **c** Morphology status

2.2 Sperm Vitality

Sperm vitality is a reflection of live membrane intact spermatozoa. It is based on the ability of cell membrane to exclude vital stains (Eosin–Nigrosin dye) from entering the spermatozoa & permeate into its nucleus. When physically damaged or broken, the eosin Y dye is able to stain sperm. If membrane is intact, the dye is unable to do so. Vitality thus evaluates viability of cell, i.e.

Live (Viable) - Unstained

Dead (Non-viable) - Stained

status of live / dead along with morphometric parameters of human sperm head. Major steps explained below.

- Corrects for intensity fluctuations by normalizing the images of a stack to the same mean intensity using bleach correction macro;
- Enhancement of images so that objects can be extracted from the background;
- Sperm heads will be extracted using head extraction algorithm, thresholding the image at different stage;
- Compute total area and actual area of sperm head using thresholding the image at different stage;
- Calculation of morphometric parameters;

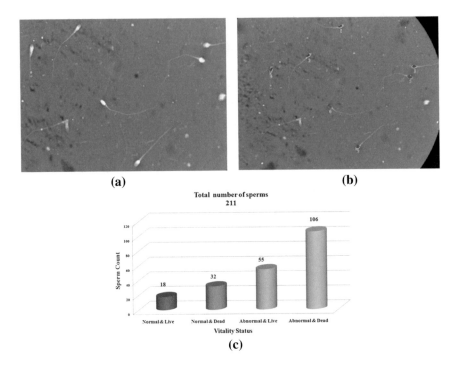

Fig. 2 **a** Input image stack **b** Output image stack **c** Status of vitality with morphology

- Apply boundary conditions on morphometric parameters as per WHO guidelines and classify normal/abnormal form;
- Compute average area, ellipticity, roundness, circularity of sperm head and classify cell live/dead.

Results

Figure 2a indicates image stack which is taken from digital microscope camera ProgRes-CT3 attached to Olympus BX-41 system microscope under 100× magnifications in the bright field mode. Figure 2b describes output image with labels which classify the sperm with colors. Red color sperm indicates dead and pink color sperm indicates live. Figure 2c points out the status of morphology and status of live/dead which is automatically generated using vitality algorithm.

2.3   Sperm Motility

Assisted reproductive techniques like in vitro fertilization (IVF), Intracytoplasmic sperm injection (ICSI), and intracytoplasmic morphologically selected sperm injection (IMSI), motile sperm is necessary to maximize the chances of fertilization. Using

Table 1 Standard values of sperm motility as per WHO manual

Motility values in micrometer/s

S.No.	Motility grade	Progression velocity	Motility status
1	Grade-A	≥ 25	Rapid progressive
2	Grade-B	$> 5 < 25$	Slow progressive
3	Grade-C	< 5	Non-progressive
4	Grade-D	0	Immotile

sperm motility analysis, we can achieve fertilization with very few spermatozoa. By combining good sperm preparation techniques with sperm motility analysis, it is possible to optimize the results as well as number of opportunities for conception [3]. Motile activity has compensating factor for low sperm counts [3]. The two aspects of motility, percent of active cells and quality, are associated with emergence of motility as a dominant factor, but the quality of motility seems to be more important than the percentage of active cells.

Table 1 shows classification of sperm motility as per their speed per second. Normal sperm exhibits motility ranging from fast progressive to slow progressive. The sperms move at a speed of 25 μ/s or 3 mm/h. A normal ejaculate will have at least 40% of progressive and non-progressive sperm. When the semen sample has less than 32% progressively motile sperm and less than 8% non-progressive sperm, the semen sample is considered to be abnormal. To examine sperm motility and its classification as per the speed per second, major steps explained below.

- Acquisition of images sequence;
- Convert in to uncompressed AVI format using FFMPEG tool;
- RGB to gray conversion, resize, calibration in μm;
- Execute sperm track plugin;
- Compute sperm motility parameters;
- Display sperm motility classification and track of each sperm with labels.

Results

Figure 3a indicates input video which is taken from digital microscope camera ProgRes-CT3 attached to Olympus BX-41 system microscope under 600× magnifications in the bright field mode. Figure 3b describes output video having sperm with labels. Figure 3c points out the sperm motility grade which is generated automatically using motility algorithm. Figure 3d indicates trajectories of sperms between the frames.

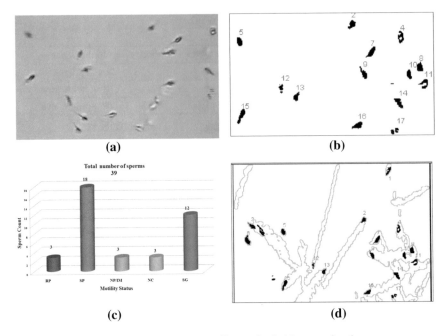

Fig. 3 **a** Input video, **b** out video, **c** sperm motility grade, **d** trajectory of each sperm

3 Conclusion

The method implemented in Sects. 2.1 and 2.2 has allowed for an exact determination
of the parameters of sperm morphology. However, the morphology and vitality seen
with the microscope is not the true morphology of living sperm, but image we create.
The image comprises a number of factors: seminal plasma, smearing technique,
fixation, staining, mounting, and the quality of microscope used. Because of all above
factors, we computed normal morphometric values of sperm head at 75% confidence
interval as compared to normal values mentioned in WHO manual. With standardized
and quality control methods, we can minimize technique dependent sources of errors
and focus our efforts on classifications of variations in sperm morphology, vitality,
and motility. The above methods seem to be a valuable tool for technicians with
substantial data for more objective diagnosis, less reliant on his previous practical
experience.

Acknowledgements We are thankful to third author of this paper Dr.Vijaykumar C.N, Cheif
Embryologist, Freelance Consultant, Belagavi, Karanataka, for the help to get human sperm samples. We have used leftover samples which have arrived to clinics/labs for analysis. Informed consent
by patients to use the sample for reasearch purposes was obtained.

References

1. Organisation WH (1999) WHO laboratory manual for the examination of human semen and sperm-cervical mucus interaction. Cambridge University Press
2. Organization WH et al (2010) Who laboratory manual for the examination and processing of human semen
3. MacLeod J, Gold RZ (Sep 1951) The male factor in fertility and infertility: II. Spermatozoon counts in 1000 men of known fertility and in 1000 cases of infertile marriage. J Urol 66(3):436–449

Market Data Analysis by Using Support Vector Machine Learning Technique

Raghavendra Reddy and Gopal K. Shyam

Abstract Stock market has been the goal of numerous market analyses which proves to be a challenge to characterize and appreciate the trends. This paper proposes a system of stock price trend prediction. It utilizes a supervised approach to cut raw transaction data of each stock into many products with a predefined value and classifies them into four main classes (open, high, low and close) according to their close prices. It contributes to analysis for market stock data classification that helps to study data in terms of stock rates where the polarity is often found to be highly unstructured. Once the data is preprocessed, then extraction of the feature from the dataset, which is meaningful, is known as feature vector. Then, the selected feature vector list is applied to the machine learning-based classification technique such as support vector machine (SVM) to predict the stock market data. The proposed method is evaluated on two publicly available markets to illustrate its effectiveness. Finally, the performance of the classifier is measured in terms of prediction accuracy.

Keywords Data cleaning · Term Frequency–Inverse document frequency (tf-idf) · Extraction technique · Support vector machine (SVM)

1 Introduction

The prediction of stock prices often poses a lot of problems due to its volatile nature. Stock markets' variation is affected by many macro-economical factors, such as firms, policies, bank rates, inverters' expectations, commodity price index and movements of other stock data [1]. Stock market is useful to provide a structured and regulated exchange where investors can safely buy and sell equity shares. We need stock market to regulate and control the transactions which take place on day-to-day activities. We

R. Reddy (✉) · G. K. Shyam
School of C and IT, REVA University, Bangalore, India
e-mail: reddycs2004@gmail.com

G. K. Shyam
e-mail: gopalkrishnashyam@reva.edu.in

© Springer Nature Singapore Pte Ltd. 2019
N. Chaki et al. (eds.), *Proceedings of International Conference on Computational Intelligence and Data Engineering*, Lecture Notes on Data Engineering and Communications Technologies 28, https://doi.org/10.1007/978-981-13-6459-4_3

19

need a common platform where it can be exchanged with ease. The best stock market research tool consolidates the necessary company information in a quick, easy-to-use format. The main goal of this paper is to analyse the stock market data with the supervised machine learning technique (SVM).

The remainder of this paper is organized as follows. In Sect. 2, the literature survey is discussed. In Sect. 3, the methodology of our proposed system is introduced. Section 4 gives the experimental results. Finally, in Sect. 5 conclusion of the proposed system is presented.

2 Literature Survey

Patel et al. [2] propose the predicting of stock market index by using a fusion model of machine learning algorithms. This model is focused on predicting future prices of stock market index. CNX Nifty and S&P Bombay Stock Exchange (BSE) Sensex indices are selected to test the data. The result is obtained with the help of ten years of historical data. The prediction is made for 1–10, 15 and 30 days in advance. In the next stage, they use support vector regression (SVR), artificial neural network (ANN) and random forest (RF) techniques. The results of SVR-ANN, SVR-SVR and SVR-RF are compared with each other.

Geetika et al. [3] propose a sentiment study of Twitter information using semantic analysis and machine learning approaches. They introduce a method for classifying the product reviews and sentence on Twitter data. Naive Bayes method is used in the system, and it provides the best result than maximum entropy (ME) and SVM. This accuracy performance improves when semantic analysis such as WordNet [3] is followed. The process is then extended to the WordNet feature for summarisation of reviews. This provides the best visualization of content which helps large number of users.

Wu et al. [4] propose a big data from the data mining perspective. They present the Huge Autonomous sources Complex Evolving associations (HACE) theorem. They analyse the challenging issues in the data-driven method as well as in big data. The proposed system provides the most accurate and most relevant social sensing feedback to understand data at real time. Further, they simulate the participation of public audience in production of the data circle for economical and social events.

Maldonado et al. [5] propose an analysis of advance conjoint process by using attributes' selection from SVM. Conjoint analysis is used to identify the consumer preferences on potential services or products. Further, the authors identify the consumer preference and the most relevant features. Initially, they focus on the SVM formulation. This in turn proves to possess significant predictive ability in marketing contexts and operation management. The experimental results depict that the system has a good fit and predictive accuracy.

Da Silva et al. [6] propose the concept of tweet data analysis using classifier ensembles and lexicons. They compare a promising strategy for the presence of tweet which is a bag of words. Feature hashing is helpful for users who can use the

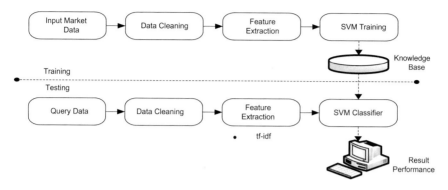

Fig. 1 Proposed system for market data analysis

analysis to search for the products and companies with public sentiment brands and for other applications. In further works, they plan to study the neural tweets where the database is large with analogue field having different attributes.

Anjaria et al. [7] propose the opinion mining influence factor of Twitter data with the help of the supervised learning method. Social network of Twitter data provides a good platform to measure the public opinion with approximately 88% accuracy. They observe PCA which is incorporated by SVM to decrease the dimensions. The inclusion of more influential factors on personal details enhances the prediction procedure to higher prediction levels. They conclude that the behavioural analysis parameters of social media can increase the prediction accuracy with sentiment analysis. In future work, opinion mining would be performed on the basis of efficient, fast and hybrid classifier using parallel computing.

3 Methodology

Figure 1 shows the proposed system of data analysis for stock market. It is separated into two phases such as training phase and testing phase. In training phase, all the data are trained and stored in a knowledge base. Stock market data is taken as input data. The cleaned data is extracted by data extraction algorithm term frequency–inverse document frequency (tf-idf) [8–10]. And it is stored in the knowledge base. In testing phase, we are testing individual data. Here, tested features are compared with knowledge base and then given polarity (high or low) rates. Data cleaning, feature selection and classification are considered as very important steps for the proposed system.

Feature selection is used to choose the essential terms that must be suitable for the document category, elimination of redundant and misleading terms. tf-idf method is used to extract attributes for data analysis. This tf-idf method expresses a weight of

Fig. 2 Hyper-planes of
SVM

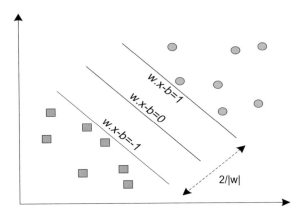

features in terms of product frequency, and inverse document frequency is shown in
Eq. (1)

$$W_g = \text{tf}_{ij} * \text{idf} = \text{tf}_{ij} * \log\left(\frac{N}{n} + 0.01\right) \tag{1}$$

where w_{ij} depicts the weight of features t_j to the category c_i. Tf_{ij} depicts the occurrences of features in terms of t_j in all documents. N depicts the total number of the
documents which contain t. Additional 0.01 prevents the weight when N is equal to
n. Equation (2) shows the working of tf-idf method.

$$w_d = f_{w,d} * \log\left(\frac{|D|}{f_{w,D}}\right) \tag{2}$$

where D depicts document collection, w depicts word, and individual document (d)
belongs to D, $f_{w,d}$ is equivalent to number of w times which can show in d, and
$|D|$ depicts size of corpus f_w, D equivalent to amount of documents in w. There are
different conditions which can occur for each word and depend on $f_{w,d}$, $|D|$ and $f_{w,D}$
values.

In this paper, the classification of the data is achieved by SVM [11, 12]. We plot
individual data from point in n-dimensional space with each feature value being
a value of particular coordinate. Then, we can perform recognition to find out the
hyper-plane which can differentiate two classes properly.

Figure 2 shows the maximum margin of hyper-planes with samples for SVM training from two classes. This SVM falls into an interest of two areas such as large margin
classifier and kernel approaches. Consider large margin or distance between two parallel hyper-planes which have a better normalization error of classifier. Assume that
the data points are $\{(x_1, y_1), (x_2, y_2), (x_3, y_3), \ldots, (x_n, y_n)\}$.

$y_n = 1/-1$ is a constant and denotes a point belongs to x_n, where n depicts number of samples and x_n depicts the p-dimensional of real vector. To view the training data, i.e. separating the hyper-planes, it is shown in Eq. 3

$$w \cdot x + b = 0 \tag{3}$$

where w depicts p-dimensional vector and b depicts scalar. The w vector points perpendicular to separating the hyper-plane. Adding offset parameter b permits to increase a margin. If b is absent, then the hyper-plane is forced to pass from the origin and restricts the solutions. The parallel hyper-planes are described in Eqs. 4 and 5.

$$w \cdot x + b = 1 \tag{4}$$

$$w \cdot x + b = -1 \tag{5}$$

If training data is separated linearly, we select the hyper-planes and then try to maximize their distance. To find out the distance between two hyper-planes, we must minimize the $|w|$. To ensure the data points either $w \cdot x_i - b \geq 1$ or $w \cdot x_i - b \leq -1$. This can be written in Eq. 6

$$y_i (w \cdot x_i - b) \geq 1, \quad 1 \leq i \leq n \tag{6}$$

Figure 3 shows the flow chart of proposed system. Initially, the system will pre-process an input data and get tokenized. Next step is to remove the stop words and create the feature vector to extract the data. Features get stored by tf-idf technique and applied to the SVM classifier. The data analysis of stock market system steps is shown in Algorithm 1.

Algorithm 1: Data Analysis

Input: Stock Market Data

Output: Identifying the Positive or Negative Rates of Data

    Start

Step 1: *Data Pre-Processing the Market Data*

        Remove missing values in column

        Remove missing values in row

        Eliminate noisy data

        Remove the Natural Common Text

Step 2: *Get Feature Vector*

        If (data in stopwords)

        continue

        else

        append the file

        Return Feature Vector

Fig. 3 Flow chart diagram
of the proposed system

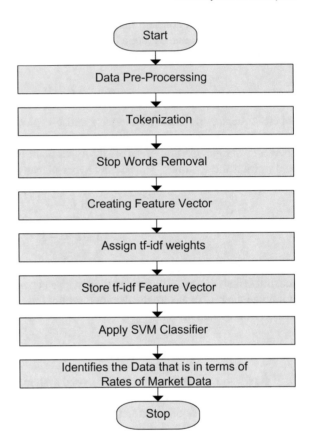

Step 3: Extract Features from Vector List

   Assign the weight to commonly used words

   Assign the weight to words that are not more frequently used

   tf * abs(idf)

   Return Features

Step 4: SVM Training

Step 5: Test the Data

Step 6: Recognize the Positive or Negative Rate

   End

4 Experimental Result

The results are obtained through support vector machine and their relative perfor-
mances are compared on accuracy parameters. Tables 1 and 2 represent the two different stock data in the year of 2017. They are FAX (Aberdeen Asia-Pacific Income

Fund, Inc.) and ACU (Acme United Corporation). In this table, prices are represented in the form of open, high, low and close price of the stock data. It depicts for each month from March 2017 to December 2017. Table 3 depicts the comparison table of existing and the proposed system for accuracy parameter. This can be measured in percentage and is computed as Eq. 7.

T_p represents the true positive rate, T_n represents true negative rate, F_p represents false positive, and F_n represents false negative rate.

$$\text{Accuracy} = \frac{T_p + T_n}{T_p + T_n + F_p + F_n} \qquad (7)$$

Figure 4 shows the comparison of the existing system with the proposed system. While comparison, we consider the combination of Naive Bayes and maximum entropy, WordNet and combination of tf-idf and SVM. Anjaria et al. [7] propose

Table 1 FAX (Aberdeen Asia-Pacific Income Fund, Inc.) stock price in 2015

Month	Open price	High price	Low price	Close price
Mar	5.45	5.47	5.38	5.38
Apr	5.58	5.59	5.53	5.55
May	5.36	5.37	5.3	5.3
Jun	4.93	4.98	4.92	4.96
Jul	4.7	4.75	4.7	4.73
Aug	4.52	4.53	4.5	4.5
Sep	4.48	4.5	4.46	4.5
Oct	4.79	4.8	4.74	4.75
Nov	4.55	4.55	4.5	4.5
Dec	4.56	4.6	4.55	4.57

Table 2 ACU (Acme United Corporation) stock price in 2015

Month	Open price	High price	Low price	Close price
Mar	18.28	18.28	18.27	18.27
Apr	18.49	18.5	18.45	18.45
May	17.5	18	17.5	18
Jun	18.1	18.25	18.1	18.1
Jul	18.25	18.39	18.1	18.27
Aug	17.25	17.35	17.25	17.3
Sep	17.23	17.23	17.23	17.23
Oct	16.48	16.7	16.3	16.35
Nov	17.12	17.12	17.12	17.12
Dec	17.5	17.5	17	17.4

Table 3 Comparison table of existing and proposed system for accuracy

S. No	Author	Methods	Accuracy (%)
1	Anjaria et al. [7]	Naive Bayes, ME	88
2	Geetika et al. [3]	WordNet	89.9
3	Proposed system	tf-idf, SVM	93

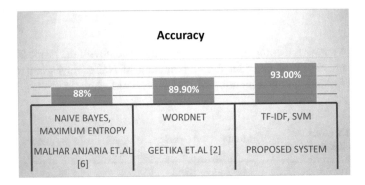

Fig. 4 Result

that, by using Naive Bayes and maximum entropy methodology, they can achieve 88% accuracy. Geetika et al. [3] propose that, by using WordNet methodology, they can achieve 89.90% accuracy. But by using the combination of tf-idf and SVM methodology, our proposed work achieves 93% accuracy.

5   Conclusion

For small start-up investment industries, due to limited funds, it is impossible to trade in the stock market regularly. We considered the stock market data to predict the rates whether they are high or low. This paper proposes a novel system which describes the market data analysis scheme by using supervised learning algorithm SVM and tf-idf feature vectors. This paper gives a better accuracy rate compared to Naive Bayes and maximum entropy [7] and WordNet [3] approach.

References

1. Rizvi SAR, Dewandaru G, Bacha OI, Masih M (2014) An analysis of stock market efficiency: developed vs Islamic stock markets using MF-DFA. Phys A: Stat Mech Appl 407:86–99 (Elsevier)

2. Patel J, Shah S, Thakkar P, Kotecha K (2015) Predicting stock market index using fusion of machine learning techniques. Expert Syst Appl 42(4):2162–2172 (Elsevier)
3. Gautam G, Yadav D (2014) Sentiment analysis of twitter data using machine learning approaches and semantic analysis. In: Contemporary Computing (IC3), IEEE, pp 437–442
4. Wu X, Zhu X, Wu G-Q, Ding W (2014) Data mining with big data. IEEE Trans Knowl Data Eng 26(1):97–107
5. Maldonado S, Montoya R, Weber R (2015) Advanced conjoint analysis using feature selection via support vector machines. Eur J Oper Res 241(2):564–574 (Elsevier)
6. Da Silva, Nadia FF, Hruschka ER, Hruschka ER (2014) Tweet sentiment analysis with classifier ensembles. Decis Support Syst 66:170–179
7. Anjaria M, Guddeti RMR (2014) Influence factor based opinion mining of twitter data using supervised learning. In: Communication Systems and Networks (COMSNETS), IEEE, pp 1–8
8. Liu M, Yang J (2012) An improvement of TFIDF weighting in text categorization. In: International proceedings of computer science and information technology, pp 44–47
9. Ramos J (2003) Using tf-idf to determine word relevance in document queries. In: Proceedings of the first instructional conference on machine learning, vol 242, pp 133–142
10. Yi J, Yang G, Wan J (2016) Category discrimination based feature selection algorithm in Chinese text classification. J Inf Sci Eng 32(5):1145–1159
11. Barnett A, Santokhi J, Simpson M, Smart NP, Stainton-Bygrave C, Vivek S, Waller (2017) Image classification using non-linear support vector machines on encrypted data. In: IACR, p 857
12. Bhavsar H, Panchal MH (2012) A review on support vector machine for data classification. Int J Adv Res Comput Eng Technol (IJARCET) 1(10):185

Prediction of Term Labor Using Wavelet Analysis of Uterine Magnetomyography Signals

T. Ananda Babu and P. Rajesh Kumar

Abstract The objective of the research is to predict the term labor by analyzing the uterine magnetomyography signals of term labor. Previous work limited to the detection of uterine contractions by extracting the features. To date, the existing research for labor prediction did not achieve high discrimination accuracy that a clinical application requires. Discrete wavelet transform is used in the research to decompose the signals. Variance, standard deviation, waveform length, energy, and entropy of wavelet coefficients are extracted from the signals of the Physionet mmgdb database. The features were divided into labor and antepartum groups. Five different classifiers were implemented to discern the two groups. Wavelet coefficient features combined with the random subspace ensemble classifier produced a powerful tool for labor assessment.

Keywords Labor prediction · Magnetomyography (MMG) · Physionet mmgdb database · Discrete wavelet transform (DWT) · Random subspace ensemble classifier

1 Introduction

Labor prediction depends on the analysis as well as characterization of the uterine electrical activity [1, 2]. Different methods like tocography (TOCO), electro hysterogram/electromyogram (EHG/EMG), intrauterine pressure catheter (IUPC), and magnetomyogram (MMG) are used for this purpose. Each method has some own merits and demerits, but are very helpful to understand the parturition process [3–5]. Nowadays, two methods are mostly used to report the physiological activity of the uterine contractions, electromyography (EMG), and magnetomyography (MMG).

T. Ananda Babu (✉) · P. Rajesh Kumar
Department of ECE, AUCE (A), Andhra University, Visakhapatnam 530003, India
e-mail: anand.mits11@gmail.com

P. Rajesh Kumar
e-mail: rajeshauce@gmail.com

© Springer Nature Singapore Pte Ltd. 2019
N. Chaki et al. (eds.), *Proceedings of International Conference on Computational Intelligence and Data Engineering*, Lecture Notes on Data Engineering and Communications Technologies 28, https://doi.org/10.1007/978-981-13-6459-4_4

Prediction of preterm delivery is achieved by the analysis of power spectrum [6] and conduction velocity of the action potentials [7].

Eswaran et al. [3] reported the uterine activity recordings (uterine MMG) by using the superconducting quantum interference device array for reproductive assessment (SARA) system. It is a noninvasive method that measures the regions of localized activation. The MMG signals does not dependent on tissue conductivity and collected outside the skin without making any electrical contact. The uterine magnetic activity analyzed by synchronization index will be helpful for the labor prediction [8]. The uterine MMG signal was segmented and the segments were classified (K-means clustering algorithm) for the detection of uterine contractions [9]. Frequency domain features are computed from the Fourier analysis [10] to differentiate labor group from antepartum group. Hilbert amplitude used as the feature to detect uterine contractions while the actual signal was decomposed by wavelet transform [11]. The sensor space partitioned to different quadrants, center of gravity, conduction velocity, and percent of active sensors were computed for characterizing the propagation of magnetomyography signals [12, 13]. However, frequency and the synchrony of electrical activity are good indicators for prediction of labor [14]. In [15], Hilbert phase approach is employed for the active labor prediction. Prediction of labor is attempted using frequency domain features [2, 5]. Diagnosis of labor is achieved after dividing the patients as labor and antepartum groups and comparing their features [6, 10]. Maner et al. [16] performed the classification for term patients with ANN classifier. Recently, Ren et al. [17] proved that the prediction of preterm labor can be improved with empirical mode decomposition (EMD) analysis.

So much work exists, for labor prediction using different classifiers, but they focused only on uterine EMG signals. The work on uterine MMG signals is limited to detection of uterine contractions by extracting the features. To the best of our knowledge, wavelet transforms are applied for the MMG signal classification for the first time in this research. Term labor records from the Physionet mmgdb database divided into two groups—labor and antepartum. Labor and antepartum groups have the records of those patients delivered within 48 h and after 48 h, respectively. All the signals are decomposed to six levels by using DWT. The wavelet coefficient features like variance, standard deviation, waveform length, energy per waveform length, and entropy were extracted from each level. Five different classifiers fed with feature space for evaluating the ability of the classifiers in labor assessment.

2   Materials and Methods

2.1   Data Acquisition and Preprocessing

The MMG signal records of term labor from the Physionet mmgdb database are used for this research [18]. The magnetomyographyy (MMG) signals are recorded at University of Arkansas for Medical Sciences, USA with the 151 channel SQUID

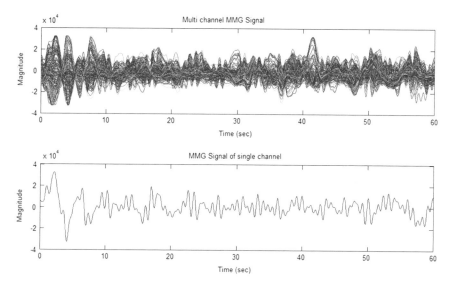

Fig. 1 MMG signals of a patient (210_37w3d) with all channels (upper trace) and a single channel (lower trace)

Array for Reproductive Assessment (SARA) systems [13]. The patient can sit comfortably and leans her abdomen on the sensor array that covers 1300 cm^2 surface area with 45×33 height and width (in cm), respectively. The records have channel ids and have the information of clinical parameters like cervical dilation, days to delivery, BMI, and gestation period. The MMG signals were digitized with the frequency 250 Hz. Then, the data is downsampled to 32 Hz and passed through a bandpass filter (0.1–1 Hz) to attenuate maternal and fetal cardiac signals. Maternal breathing is the prominent signal (0.33 Hz), which is suppressed with notch filter (0.25–0.35 Hz). Final recording was excluded with the signal segments of maternal movement. Final database contains 147/148 channels, and each signal lasts between 10 and 20 min in which the first 1-min record is represented in (Fig. 1).

The upper trace shows the multichannel signal (148 sensors), and the lower trace represents the single-channel signal. In the Fig. 1, subject name shows by ***_#GP, where *** represents the subject ID (210), and #GP refers to the gestation period in weeks and days (37w3d). The magnitude on y-axis is represented in raw units. For the physical units (pT), subtract the gain of each channel from the magnitude and divide the result with its base.

2.2 Wavelet Analysis

Wavelets are some mathematical tools which can be utilized to get the information from different physiological signals. Wavelet transforms (WT) allowed the trans-

Fig. 2 Discrete wavelet transform tree with six decomposition levels

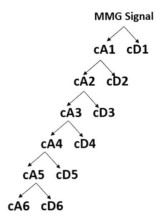

formed data to be analyzed in both time and frequency domains like signal-processing algorithms. Discrete wavelet transform (DWT) is described in terms of filter banks for signal-processing application. The MMG signal is passed through low- and high-pass filters to get approximate and detailed coefficients. This approach is nothing but the multi-resolution decomposition of the signal. The transformed coefficients present the energy distribution of MMG signal in both domains.

The approximate coefficients further divided into detailed and approximate coefficients. By choosing the mother wavelet, the coefficients of the filter banks can be regulated. The selection of mother wavelet and decomposition levels are vital issues. The difficulty lies in the decision of selecting a wavelet, and the scale that will bring the best result for a particular application. Biorthogonal wavelet sym8 from the symlets family with six decomposition levels is employed in our research. Figure 2 represents the general DWT tree with six decomposition levels. cAn and cDn in Fig. 2 represent the n-th level approximate coefficients and the detailed coefficients, respectively.

Since the MMG is a multichannel signal, a generalized one-dimensional multi-signal wavelet analysis applied for extracting the features. For a given database of MMG term records, the decomposition for the signal starts along the columns (Channel) up to six levels. Consider the each level as a feature space, the features were constructed by calculating the transformed coefficients variance, standard deviation, waveform length, energy per waveform length, and entropy. The feature space for each individual patient has seven feature vectors, since the DWT decomposes the signal to seven coefficient vectors (cA6, cD1–cD6). So, the feature matrix contains 35 (5 features × 7 feature vectors) different sample values corresponding to each channel.

2.3 Classification

The Naïve-Bayes (NB) is a simple sorting algorithm with many applications like spam mail filtering, medical diagnosis, and weather prediction. The classifier deploys the Bayes theorem to compute the class probability. It primarily depends on the assumption of class independence. In most classification problems, conditional independence is not possible. The advantage of Naïve-Bayes lies in the fact that even a modest amount of training information is sufficient for better classification. K-nearest neighbor (KNN) is a lazy classifier that can be used for many applications related to the biomedical signal classification. The test data will be classified depends on the class of the nearest neighbor. KNN classifier with Euclidean distance metric is used for our research.

Support vector machine (SVM) has a successful record in the classification of uterine physiological signals. Linear kernel, polynomial kernel, sigmoid kernel, and RBF kernel are the most commonly used kernel functions. In our work, we used the polynomial kernel function. Artificial neural networks (ANNs) are ideal in the classification of patients for the evaluation of the risk of preterm labor. ANNs consist of three different layers—input layers, hidden layers, and output layers. The hidden layers have artificial neurons (analogous-to-brain neurons) which are used to process the data. Scaled conjugate gradient backpropagation function is used in our network to update the weight and bias values.

The random subspace classifier (RSC) is an ensemble classification method which is suitable for a large number of features. This method can be used with many classifiers like NB, SVM, and KNN and others as base classifiers. The ensemble decision of the class label depends on either majority voting or probabilities of base classifiers.

Random subspace classifier implemented in our research as follows:

a. The number of training vectors and the number of features in each vector are n and d
b. l is the number of base classifiers
c. d_i is the number of predictors for each classifier
d. A dataset created by taking d_i features from d with replacement to train each classifier
e. Test data applied to the ensemble model
f. The outputs from the classifiers combined by majority voting to decide the class of test data.

Classification is the identification of similar data segments. If the class of data is known, then it is termed as supervised learning. Since the supervised algorithms are sensitive to the division of data, our classifiers were trained by 2/3 of data while the remaining is the test dataset. We investigate the suitability of the classifiers for classifying the term records into labor vs antepartum groups to predict the term labor.

Table 1 Confusion matrices for Naïve-Bayes, ANN and KNN (top row), SVM, and random subspace (bottom row) classifiers

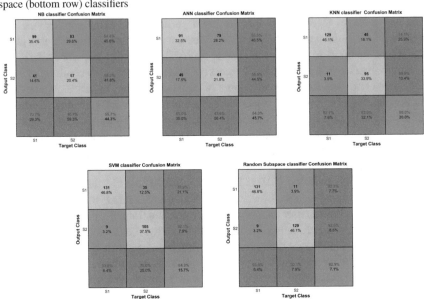

3 Results and Discussion

The classifier performance can be evaluated by observing the confusion matrices. Confusion matrix, a two-dimensional table with each row and each column represents different classes. Each element in the matrix represents the total test vectors for which the actual and predicted classes are in the row and in the column, respectively. Table 1 displays the confusion matrices of present work for the classification results of the Naïve-Bayes, KNN, SVM, ANN, and random subspace classifiers.

The performance metrics from the confusion matrices were defined as follows. Accuracy measures the ratio of both labor and non-labor signals that are correctly acknowledged as labor and antepartum. Precision or positive predictive value gives the proportion of labor signals that are identified from both labor and non-labor signals. False positive rate (FPrate) is the ratio of misclassified labor signals to the total number of signals that are not predicted as labor class. The accuracy, precision, and FPrate values presented in Table 2 for the Naïve-Bayes, KNN, SVM, ANN, and random subspace classifiers.

From the confusion matrices (Table 1) of Naïve-Bayes and ANN classifiers, we observed that the two-thirds of labor signals are correctly classified. Those classifiers performed poorly in predicting non-labor signals which results the classification accuracies 55.7143% and 54.2857%, respectively. The KNN classifier is better to classify the labor signals and gives an accuracy of 80% despite the misclassification of non-labor signals (one-third of the data). The SVM classifier performance is

Table 2 Classification results of Naïve-Bayes, ANN, KNN, SVM, and random subspace classifiers

Classifier	Accuracy (%)	Precision (%)	FPrate
Naïve-Bayes	55.7143	56.2794	0.4372
ANN	54.2857	54.2857	0.4571
KNN	80.0000	81.8803	0.1812
SVM	84.2857	85.5105	0.0395
Random subspace	92.8571	92.8922	0.0711

better than the KNN classifier (84.2857%). The random subspace ensemble classifier wins the race with lesser misclassification of non-labor signals with an accuracy of 92.8571%. The precision values of the Naïve-Bayes, ANN, KNN, SVM, and random subspace classifiers are 56.2794%, 54.2857%, 81.8803%, 85.5105%, and 92.8922%, respectively. It confirms the random subspace classifier superiority in labor prediction among the other classifiers. The FPrates of the Naïve-Bayes and ANN classifiers are 0.4372 and 0.4571, respectively, while that of KNN classifier is 0.1812. The SVM classifier FPrate is 0.0395 is smaller than the random subspace classifiers FPrate 0.0711. It confirms the classifiers' ability to correctly identify the labor signals than non-labor signals.

Various linear and nonlinear signal-processing techniques employed for labor prediction in literature. Lucovnik et al. [6] considering the labor and antepartum groups with a 7-day threshold. In [17], the empirical mode decomposition features are used with six different classifiers for the labor prediction. Maner et al. [16] classified the term/preterm labor and non-labor groups with reasonable accuracies. They employed the frequency domain features with ANN classifiers. In most of the previous studies, the researchers focused only on uterine EMG signal classification. MMG signal analysis is limited to a single-channel observation, based on RMS and zero crossings as features in [9], Fourier analysis and wavelet analysis in [10, 11, 15]. In our research, we extracted the features from wavelet coefficients of all channels which provide the information in both time domain and frequency domain. We achieved clinically acceptable discrimination accuracy. The maternal factors like gestation age (GA), body mass index (BMI), and cervical dilation/effacement which will affect the onset of labor were not considered for this study. The patients delivered within one and two days may have same maternal factors which imposes a limitation to our study. Our study is limited to only 24 subjects; we may extend the proposed method to large dataset and preterm labor records. In future, we may think about wavelet packet decomposition for feature extraction.

4 Conclusion

The prediction of term labor through the wavelet analysis of uterine MMG signals is attempted in this work. The records were separated depends on the patients days remaining for delivery. Then, the DWT is applied to decompose the signals to six levels using sym8 as the mother wavelet. Variance, waveform length, standard deviation, energy per waveform length, and entropy were extracted from the wavelet coefficients. These features applied to different classifiers to investigate their ability in predicting labor. The random subspace ensemble classifier has the better performance with 92.8571% accuracy, 92.8922% precision, and 0.0711 FPrate among the other classifiers.

References

1. Gondry J, Duchene J, Marque C (1992) First results on uterine EMG monitoring during pregnancy. Proc Ann Int Conf IEEE EMBS 6:2609–2610
2. Eswaran H, Preissl H, Wilson JD, Murphy P, Lowery CL (2004) Prediction of labor in term and preterm pregnancies using non-invasive magnetomyographic recordings of uterine contractions. Am J of Obstet Gynecol 190(6):1598–1603
3. Eswaran H, Preissl H, Wilson JD, Murphy P, Robinson SE, Lowery CL (2002) First magnetomyographic recordings of uterine activity with spatial temporal information with a 151-channel sensor array. Am J Obstet Gynecol 187:145–151
4. Horoba K, Jezewski J, Wrobel J Graczyk S (2001) Algorithm for detection of uterine contractions from electrohysterogram. In: Proceedings. 23rd international conference IEEE EMBS, pp 461–464
5. Lucovnik M, Kuon RJ, Garfield RE (2011) Use of uterine electromyography to diagnose term and preterm labor. Acta Obstet Gynecol Scand 90(2):150–157
6. Lucovnik M, Maner WM, Chambliss LR, Blumrick R, Balducci J, Novak-Antolic Z et al (2011) Noninvasive uterine electromyography for prediction of preterm delivery. Am J of Obstet Gynecol. 204(3):228
7. Rabotti C, Mischi M, Oei SG, Bergmans JWM (2010) Noninvasive estimation of the electro hysterographic action-potential conduction velocity. IEEE Trans Biomed Eng 57(9):2178–2187
8. Ramon C, Preissl H, Murphy P, Wilson JD, Lowery CL, Eswaran H (2005) Synchronization analysis of the uterine magnetic activity during contractions. Biomed Eng Online 4:55
9. La Rosa PS, Nehorai A, Eswaran H, Lowery CL, Preissl H (2008) Detection of uterine MMG contractions using a multiple change point estimator and the k-means cluster algorithm. IEEE Trans Biomed Eng 55:453–467
10. Garfield RE, Maner WL, MacKay LB, Schlembach D, Saade GR (2005) Comparing uterine electromyography activity of antepartum patients versus term labor patients. Am J Obstet Gyn 193:23–29
11. Furdea A, Eswaran H, Wilson JD, Preissl H, Lowery CL, Govindan RB (2009) Magnetomyographic recording and identification of uterine contractions using Hilbert-wavelet transforms. Physiol Meas 30(10):1051–1060
12. Furdea A, Preissl H, Lowery CL, Eswaran H, Govindan RB (2011) Conduction velocity of the uterine contraction in serial magnetomyogram (MMG) data: event based simulation and validation. In: 2011 conference proceedings IEEE EMBS, pp 6025–6028
13. Escalona-Vargas D, Govindan RB, Furdea A, Murphy P, Lowery CL, Eswaran H (2015) Characterizing the propagation of uterine electrophysiological signals recorded with a multi sensor

abdominal array in term pregnancies. PLoS ONE 10(10):e0140894. https://doi.org/10.1371/journal.pone.0140894

14. Govindan RB, Siegel E, Mckelvey S, Murphy P, Lowery CL, Eswaran H (2015) Tracking the changes in synchrony of the electrophysiological activity as the uterus approaches labor using magnetomyographic technique. Reprod Sci 22(5):595–601

15. Govindan RB, Vairavan S, Furdea A, Murphy P, Preissl H (2010) Decrement of uterine myometrial burst duration as a correlate to active labor: a Hilbert phase approach. In: 32nd Annual international conference of the IEEE EMBS

16. Maner WL, Garfield RE (2007) Identification of human term and preterm labor using artificial neural networks on uterine electromyography data. Ann Biomed Eng 35(3):465–473

17. Ren P, Yao S, Li J, Valdes-Sosa PA, Kendrick KM (2015) Improved prediction of preterm delivery using empirical mode decomposition analysis of uterine electromyography signals. PLoS ONE 10(7):e0132116. https://doi.org/10.1371/journal.pone.0132116

18. Goldberger AL, Amaral LA, Glass L, Hausdorff JM, Ivanov PC, Mark RG, Mietus JE, Moody GB, Peng C-K, Stanley HE (2000) PhysioBank, PhysioToolkit, and PhysioNet: components of a new research resource for complex physiologic signals. Circulation 101(23), e215–e220[circulation electronic pages; http://circ.ahajournals.org/cgi/content/full/101/23/e215

IoT-Based Portable Health Scrutinization

Akhil Menon, Sonali Khobragade, Onkar Mhatre and Sakshi Somani

Abstract Today, the healthcare environment has become more technology-oriented. Heart patients are facing a problem of unexpected death because of lack of medical care given to the patient at right time. Thus, patient monitoring is essential to care in operating and emergency rooms, as well as intensive and critical care settings. We propose an IoT-based patient monitoring system which provides an improved health care to people in a more economic and significant sociable manner. In this system, a patient carries a hardware having sensors and android phone application; the sensors sense the different parameters of the patient, and this data is transferred to the cloud of ThingSpeak. System has the cloud record which stores all statistics about patient's health, and the doctors can prescribe medicine using this information. Continuous monitoring and recording of multiple patients are possible at the same time using our phone application—ProHealth.

Keywords Health care · Patient monitoring · Sensor · Application · ThingSpeak

1 Introduction

Health care is the most vital concern of every country in the world as it contributes to a significant part of a country's economy. According to the World Health Organization (WHO), an efficient healthcare system requires robust and well-maintained healthcare facilities. Hospitals always need better supervision. There is no facility to

A. Menon (✉) · S. Khobragade · O. Mhatre · S. Somani
Ramrao Adik Institute of Technology, Sector 7, Phase I, Nerul, Navi Mumbai 400708, India
e-mail: akhilmenon96@gmail.com

S. Khobragade
e-mail: sonu.khobragadeo42@gmail.com

O. Mhatre
e-mail: onkarmhatre2@gmail.com

S. Somani
e-mail: sssomanisakshi@gmail.com

© Springer Nature Singapore Pte Ltd. 2019
N. Chaki et al. (eds.), *Proceedings of International Conference on Computational Intelligence and Data Engineering*, Lecture Notes on Data Engineering and Communications Technologies 28, https://doi.org/10.1007/978-981-13-6459-4_5

check the constraints of patients with critical illness when they return to home. And hence, there is a possibility that the disease may return again, or the patient's health may become worse. Studies from various sources have shown that 30% of patients who are readmitted at least once within 90 days suffer from a discharge diagnosis of heart failure with readmission rates ranging from 25 to 54% within 3–6 months. In response to these needs, it is important to develop a remote patient health monitoring system for post-medical discharge [1].

Today, technology plays an important in every aspect of our lives. Out of all the industries, technology can be splendidly applied to health care for better diagnosis and improved patient care. Internet of things (IoT), a branch of embedded systems is a network of devices through software and sensors to enable them and exchange data. It has the advantage of connecting a huge number of devices at the same time. It also includes data storage and "predictive interaction" features which makes IoT unique. IoT is very widely used for consumer usage to ease their lives. It is used in remote health monitoring of the patients which can range from blood pressure to heart rate monitoring with emergency alert mechanism [2, 3].

2 Proposed Work

Remote health monitoring is a rapidly evolving perception since the health of the patient becomes a vital concern post-heart surgery or any precarious disease. Thus, continuous real-time intensive care is attempted to provide to the patient. The main aim of the system is to deliver an individual's health nursing by keeping a check on the body temperature, heart rate, and ECG waveform. An android application has been developed to view the individual's health status from time to time.

The interconnection between different modules is shown. The patient's data (body temperature, heart rate, and heart activity) are frequently measured by monitoring sensors that relate to the patient's body. The other end of the sensors is connected to Raspberry Pi 3. The data sensed and calculated by the sensors are stored in the Raspberry Pi 3 and continuously sent to a server/cloud on ThingSpeak, an IoT platform, as well. The period of sending or the intervals of receiving and storing the data can be set. The cloud data can be retrieved and checked in the android phone application—ProHealth, by the patient as well as the doctor. Pi compares the data with the normal range of values for an individual. If the values exceed the normal range of the measuring parameter, an emergency response is sent to doctors, close relatives, and the hospital. All the values are stored in ThingSpeak, and the most recent value is displayed using an android application. Patients and doctors are given unique username and password to view their health records. This ensures the security of the records and no need to handle any files. The doctor along with their username/login credentials can log in and see the patient's health parameters. Doctors can see all previous records of a patient and suggest medicines and changes in prescription.

3 Implementation

Raspberry Pi does not support analog inputs. So, in order to connect analog sensors, we make use of ADC MCP 3208. It is a 16-pin IC which is capable of providing eight channels simultaneously. MCP 3208 sends data to Pi using serial port interface (SPI) protocol. For this, the four control pins, viz. CLK is connected to GPIO 11 (Pin 23) of Pi, Dout is connected to GPIO 09 (Pin 21) of Pi, Din is connected to GPIO 10 (Pin 19) of Pi, and CS/SHDN is connected to GPIO 08 (Pin 20) of Pi. The ECG is connected to Pi, and the probes are connected to the patient's body in any of the 12 lead configurations.

On booting the system, the Raspberry Pi takes the value from the sensors through MCP3208. It then sends the data to cloud every 10 min and displays the same on an LCD. The android application which is prepared using Android Studio also displays the value of health parameters for the respective patients. The doctor also looks after the analysis and prescribes the steps to be taken. The threshold value for body temperature of the patient is set to 100 K, and the threshold value for heartbeat is set to 90 bpm. When the sensors sense a value beyond these values gives an alert notification to the doctor. The doctor then takes the necessary steps. The statistics of the parameters are also displayed on the cloud. The flowchart in Fig. 1 shows the flow of the system.

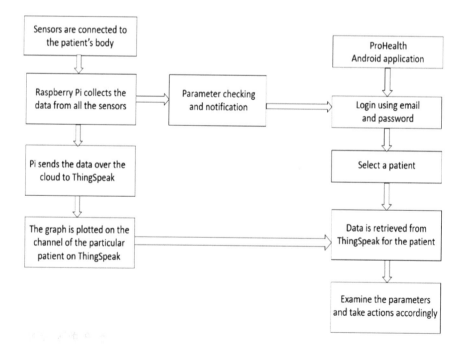

Fig. 1 Flowchart

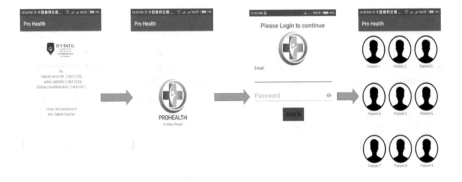

Fig. 2 Android application flow

3.1 Android Application Development

Different activities are created for each screen of the app. First, an activity for showing the name of the author and the institute was created. The second activity was created showing the logo of the application. The third activity includes the login credentials. The application takes input from the user and gives access further if the input is correct. The next activity shows the list of patients. The doctor can select any one patient and move forward. Further, the values of the parameters are retrieved and displayed for the particular patient in available Internet connection (Fig. 2).

4 Result

4.1 Heart Rate and Body Temperature

The health parameters of a patient have been noted and tabulated as shown. The normal range of heart rate for adults is 60 to 100 while for the normal body temperature is 98.6 °F (37 °C). The heart rate for a normal person increases during any physical activity by the individual and can increase up to 190 bpm. Table 1 shows the body temperature and bpm for patient 1. The graph is plotted values v/s time on the ThingSpeak platform and retrieved in the application. Body temperature values, which are measured in centigrade, are converted to Fahrenheit, whereas the heart rate is measured in beats per minute (bpm) (Fig. 3).

Observations for patient 1: The heart rate of the patient is 110 highest and then decreases gradually with time. The patient might have done some physical activity and then come to rest. Finally, the reading of 62 bpm is recorded which indicates that the patient is doing zero body movement when he is asleep. At the same time,

Table 1 Readings for patient 1

S. No.	Body temperature (°F)	Heart rate (bpm)
1	100	110
2	98.2	104
3	95.1	91
4	92.8	81
5	95.6	89
6	100	76
7	99.1	69
8	96.3	62

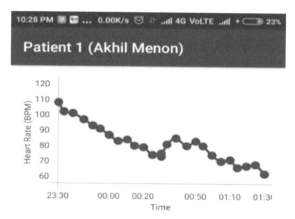

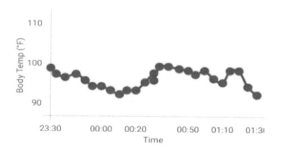

Fig. 3 Graph in the android application for patient 1

Fig. 4 Graph of ECG

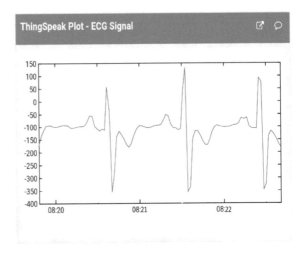

the body temperature of the patient varies from 100 °F to 92.8 °F with the average temperature being 97.13 °F which is considered normal.

4.2 ECG

The ECG reading of one of the patients is taken, and the graph is obtained (Fig. 4). From the ECG graph obtained, the following analysis has been made.

1. Heart rhythm and heart rate: The P wave present before the QRS complex has a sinus morphology and is called as a normal sinus rhythm (NSR). A sinus morphology is an upright P wave in Lead II. The distance between R-R waves is 0.9 s, whereas the distance between P-P waves is 0.10 s. Heart rate = (60/0.9) = 66.67.

2. Waves which reflect the conduction: The PQ interval is of 190 ms. The width of the QRS complex which is measured from the beginning of the Q wave to the end of the S wave is of 0.07 s (70 ms). Abnormal Q wave indicates myocardial infarction. QT interval is measured from the beginning of the QRS complex to the end T wave which is found to be 440 ms. The shortening of QT interval is a sign of hypercalcemia. All these values indicate that the patient's ECG is in normal state.

3. Electrical axis of heart: The height of R wave for second P-QRS-T wave is 240, and the height of S wave is 200. Since the height of R wave is more than the height of S wave, the ECG is normal. If the height of S wave is more than the height of R wave, then it indicates problems in the ventricles.

4. QRS complex: The width of QRS complex is 120 ms in normal state. The width of the interval in the graph is 100 ms.

5 Conclusion

An extension of the proposed work will be including a glucose-level detector sensor and also continuously detecting the movements of the patient using a motion sensor. Another extension is to include wireless sensor area networks (WSANs) to activate the drug delivery system, and muscle simulator during emergencies.

IoT-based portable health scrutinization is an effective system in carrying out real-time and long-term monitoring of patient's health. It provides an emergency rescue mechanism and alerts the hospitals and relatives using the ProHealth mobile application. The patient's previous medical history is stored on the cloud using which becomes easier for the doctors to determine the further prescriptions without the need of going through any paper files.

Acknowledgements IoT-based portable health scrutinization is oriented toward remote technology and the use of IoT for an improved system and does not focus on any kind of medical research. The results shown were examined with patient's consent and focus on how IoT works according to the proposed idea.

References

1. Mohammeda ZKA, Ahmedb ESA (2017) Internet of things applications, challenges and related future technologies. Article, Jan 2017
2. Rugierri M, Nikookar H ()Internet of Things—from research and innovation to market development. ISBN: 978-87-93102-95-8
3. Sebastian S, Jacob NR, Manmadhan Y, Anand VR, Jayashree MJ (2012) Remote patient monitoring system. Int J Distrib Parallel Syst (IJDPS) 3(5)

Automatic Evaluation of Programming Assignments Using Information Retrieval Techniques

Md. Afzalur Rahaman and Abu Sayed Md. Latiful Hoque

Abstract Nowadays, automatic assessment is a common need for the programming courses in e-learning platform. Programming courses often have a huge number of assignments, which is much tedious and error-prone job to manually check by instructors. In this paper, we present a model for automatic evaluation of C programming assignments by using TF-IDF algorithm, which is one of the most promising methods of information retrieval system. For scoring and ranking document, a combination of TF-IDF and cosine similarity algorithms has excellent performance. Experimental result shows that the proposed model has a good performance level.

Keywords TF-IDF · Cosine similarity · Program solution feature

1 Introduction

Computer-based education system has been much popular day by day because of upgraded technology and cheaper cost. E-learning is the most promising computer-based education system, being popular due its simplicity and user-friendly nature. Here, an interaction between learners and instructors is easier than conventional learning system. In the current education field, programming courses have launched almost in every faculty. Students most often prefer e-learning system to learn about programming courses. In the e-learning platform, e-assessment of programming assignments is a great challenge [4]. Generally, instructors have to evaluate a huge number of assignments in programming courses. It would be much time-consuming and inefficient procedure, if instructors check assignments manually. To assess a program, they have to check the correctness of every parts like variable initialization, statements, conditions, and their scopes, for each program, which is much cumbersome. To

M. A. Rahaman (✉)
IICT, BUET, Dhaka, Bangladesh
e-mail: afzalurrahaman@yahoo.com

A. S. M. Latiful Hoque
Department of CSE, BUET, Dhaka, Bangladesh
e-mail: afzalurrahaman@yahoo.com

© Springer Nature Singapore Pte Ltd. 2019
N. Chaki et al. (eds.), *Proceedings of International Conference on Computational Intelligence and Data Engineering*, Lecture Notes on Data Engineering and Communications Technologies 28, https://doi.org/10.1007/978-981-13-6459-4_6

overcome the evaluation challenge in e-assessment, we have developed an automatic evaluation system by the combination of TF-IDF and cosine similarity algorithms.

The rest of the paper is organized as follows. In Sect. 2, we have presented the architecture of proposed system. In Sect. 3, we have analyzed algorithms, solution feature, and performance of proposed system. In Sect. 4, we have described the evaluation procedure and token error estimation of programs. In Sect. 5, we have measured model performance. Finally in Sect. 6, we have described the contribution, the limitations, and the future works.

2   Architecture of Proposed System

In this model, we have used token frequency–inverse document frequency and cosine similarity algorithms for evaluation. First, we have generated a solution corpus having every possible combination of solutions of a programming problem. Then, each program has been tokenized. After tokenization, unnecessary tokens were removed. We have measured TF-IDF weight of each token and logarithmically offset the weight. Using the TF-IDF score vectors of solution programs, we have generated a solution matrix. The cosine similarity between a submission and each solution vector of matrix was compared to measure the degree of relevance. The whole system architecture has been partitioned into two main parts: solution matrix generation and submitted program evaluation.

2.1   Solution Matrix Generation

Figure 1 shows the solution matrix generation part of the proposed system. A solution program set has been selected by passing statement coverage, branch coverage, multicondition, and path coverage. Programs were checked with the test case sets. Then, a solution matrix has been generated based on the following steps:

Program tokenization: Basically, program is the collection of different types of tokens. Hence, we sliced the entire program to generate tokens. In this stage, a variable declaration and initialization, statements, conditions, etc., were partitioned into tokens.

Stopwords and unnecessary token removal: At the time of programming, we often write user interaction message, preprocessor directives, and library functions to make the program more comprehensive and easier. These things have little contribution in programming performance evaluation. That is why we have selected a set of common words, library function names, and preprocessor directive that are often used in programming like enter, value, between, range, printf(), getch(), clrscr(), #include<stdio.h>, #include<conio.h>, etc., and added them in stopword set.

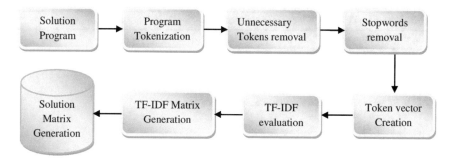

Fig. 1 Solution matrix generation

TF-IDF Evaluation: The token frequency of a token t in program p is defined as the number of times that t occurs in p. The logarithm frequency weight of token t in program p is:

$$w_{t,p} = \begin{cases} 1 + \log_{10} tf_{t,p}, & \text{if } tf_{t,p} > 0 \\ 0, & \text{otherwise} \end{cases} \tag{1}$$

Inverse document frequency is the rarity of a token in the perspective of solution corpus. The IDF weight increases as it is being rare. If a token appears in every program in corpus, its IDF is zero. If a token does not appear in corps, there would be a divide-by-zero occurrence. Therefore, we added one with denominator. We have also used logarithm in both terms to scale up the weight properly.

$$idf_{t,p} = \log_{10} \left(\frac{|Solution\,Corpus|}{1 + df_t} \right) \tag{2}$$

TF-IDF matrix generation: Finally, a solution matrix having order m x n has been generated by using the following equation. Here, m represents the length (number of tokens) of each solution and n is the number of solution.

$$tf \cdot idf_{t,p} = (1 + \log_{10} tf_{t,p}) \times (\log_{10} \left(\frac{|Solution\,Corpus|}{1 + df_t} \right)) \tag{3}$$

For solution matrix generation, we have measured the token frequency and inverse document frequency of solution set by using Algorithm 1. After vector generation of each solution, we have integrated all vectors in solution matrix. Each row of the matrix was considered as a solution vector. We used this matrix to measure correctness of a submitted program.

Algorithm 1: TF-IDF Vector Generation

Input: Solution Program Set
Output: A Solution matrix $TFIDF_{m \times n}$ Generation
for *program* $i := 1$ *to* n **do**
| Step 1:
| a. Tokenize program
| b. Remove stopwords and unnecessary tokens
| **for** *token* $i := 1$ *to* n **do**
| | Step 2:
| | a. tf $\leftarrow 1 + \log_{10} tf_{t,p}$
| | $b.idf \leftarrow \log_{10} \left(\frac{|Solution\ Corpus|}{1 + df_t} \right)$
| | $c.tfidf \leftarrow tf_{t,p} \times idf_{t,p}$
| | $d.Save\ tfidf\ weights\ in\ vector\ \mathbf{V}$
| | d. update $TFIDF_{m \times n}$ matrix by concatenating vector $\mathbf{V}$

2.2 Submission Evaluation

Figure 2 shows the evaluation part of proposed model. The submitted program has graded first by human graders. In this model, we have considered grade as correctness of program in percentage. A new submission is considered as query vector that compared with each solution vector by using cosine similarity. The maximum similarity was chosen for measuring correctness. In cosine similarity measure, we have considered each solution vector as s and submission as q.

$$\cos(\mathbf{s}, \mathbf{q}) = \frac{\mathbf{s} \times \mathbf{q}}{\|\mathbf{s}\| \|\mathbf{q}\|} = \frac{\sum_{i=1}^{n} s_i q_i}{\sqrt{\sum_{i=1}^{n} (s_i)^2} \sqrt{\sum_{i=1}^{n} (q_i)^2}} \tag{4}$$

The entire architecture has shown in Fig. 3. First, a new submission has graded by instructor. Then, program was graded by model. Instructor keeps the assessment record of model for every evaluation time to measure performance and further improvement.

During evaluation, after tokenization and removing stopwords we have measured token frequency *TF* of submission. The corresponding *IDF* values of each token were assigned from the corpus. The product of *TF* and *IDF* value was then used to generate a query vector q from the submission. Algorithm 2 has decomposed the evaluation part.

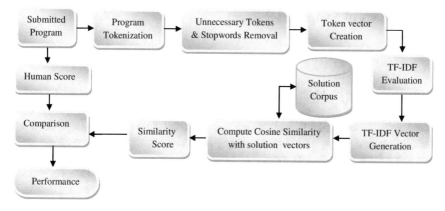

Fig. 2 Evaluation of submitted program

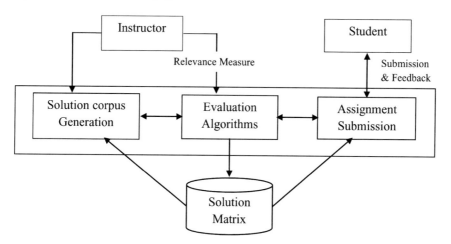

Fig. 3 Architecture of proposed model

Algorithm 2: Testing A submission

Input: Submitted program
Output: Grading submitted program
Step 1: Call TF-IDF vector Generation() to generate query vector q from the submitted program
Step 2: Compute Cosine Similarity between q and each vector of $TFIDF_{m \times n}$ matrix and choose the maximum similarity for automatic grading
Step 3: Assign grade G for the submission and measure the performance of model by comparing human expert grader

---

3 Analysis of Proposed System

3.1 Solution Feature

A C programming problem has many alternative correct solutions. Any programming problem can be written by using for, while, or do while loop. Inside a loop, there may have if, else-if condition statement. Also, a solution can be function based. For a single programming problem, there may have a solution with any possible combination of aforementioned ways. To correctly measure programming assignment, model should have every possible solution. Therefore, we have developed our model with all possible solution features. Considering every possible way, we represent the entire model into a hierarchical graph structure. A submission has to satisfy any possible route of graph to confirm loops and conditional correctness. During evaluation, after tokenization and filtering unnecessary tokens we have generated a query vector from the submitted program. Correctly written conditional statements and loops of submitted programs token were fully matched with a route of graph. This implies good performance in program writing. To demonstrate the procedure, we have considered the submission program of case 1. To measure loop and conditional tokens correctness, we have filtered the query vector as follows: [++i), for(i=2;, i< =n/2;, if(flag==0), if(n%i==0)].

The vector was fully matched with one route of hierarchical graph that ensures loop and conditional statement correctness of the submission.

3.2 Performance Analysis

To measure performance, we have considered four possible cases during evaluation. In these cases, we have highlighted the most common attitude of introductory-level students during program writing.

Case 1: Student often writes wrong token due to less understanding or memorizing programs. In this case, student does syntax error. Less caring about case sensitivity implies these types of error. This submission has six wrong tokens. After tokenization of submission, algorithm returns the query vector q as follows: [main, intn, i, print, For, (i=2, i< =n/2;, ++i), If(flag==0), break;, else, flag=0;, flag=1;, if(n%i==0), scanf("%d",&n);].

The idf values of wrong tokes were not present in solution corpus. Hence, algorithm assigned zeroes for these tokens as follows: [0, 0, 2.3, 0, 0, 0, 0.7, 0.5, 0, 0.6, 1.3, 0.5, 3.3, 1.0, 0.1].

Now, cosine similarity gives a poor matching between submitted program and correct solutions. In this case, model was successfully able to evaluate the submission.

Case 1	Case 2	Case 3

#include<stdio.h>
int *main*
{
intn, i, flag=0;
print ("Enter a positive integer");
scanf("%d",&n);
For (i=2 i<=n/2; ++i) {
if(n%i==0) {
flag=1;
break; } }
If (flag==0)
printf("%d is a prime number", n);
else
printf("%d is not a prime number", n);
return(0); }
	#include <stdio.h>
	int main() {
	  int *year;*
	printf("Enter a year: ");
	scanf("%d",&year);
	if(year%4 == 0) {
	if(year%100 == 0) {
	if (year%400 == 0)
	printf("%d is a leap year.", *year);*
	else
	printf("%d is not a leap year.", *year);* }
	else
	printf("%d is a leap year.", *year*); }
	else
	printf("%d is not a leap year.", *year);*
	return 0; }
		#include <stdio.h>
		int main() {
		int n, i, flag=0 ;
		printf("Enter a positive integer");
		scanf("%d",&n);
		for(i=2; i<=n/2; ++i) {
		if(n%i==0) {
		flag=1;
		break; } }
		if(flag==0)
		printf("%d is a prime number", n);
		else
		printf("%d is not a prime number", n);
		return(0); }
		*/*extra wrong statement:*/*
		if(st[j]>st[j+1]) {
		temp=st[j];
		st[j]=st[j+1];
		st[j+1]=temp; }

Case 2: A submitted program may not relate to correct solution. Improper problem understanding, tendency of circumventing to instructor, random guessing implies these categories wrongs. In this submission, we have found four irrelevant tokens in perspective to our solutions corpus. Hence, *idf* weight assigned to 0 for these tokens and cosine similarity returns a very poor matching alike previous case. In this case, model was also successfully able to perform fair evaluation.

Case 3: In this case, we have considered a submission having some unnecessary statements with correct solution. After tokenization, we have found 18 tokens as shown in following vectors: [`&n);`, `break;`, `else`, `flag++;`, `flag=0;`, `for(i=2;`, `i++)`, `i,`, `i<=n/2;`, `if(flag==0)`, `if(n%i==0)`, `if(st[j]>[j+1])`, `n,`, `return(0);`, `scanf("%d",`, `st[j+1]=temp;`, `st[j]=st[j+1];`, `temp=st[j];`] In this stage algorithm was ignored four tokens and return the tfidf values of 14 tokens. [0.1, 0.7, 0.5, 2.5, 0.6, 0.9, 1.0, 0.5, 3.3, 1.4, 1.0, 0, 0.3, 0.1, 0.2,0, 0, 0].

Finally, vector was fully matched with a correct solution. In this case, model has failed to perform fair evaluation.

Case 4: In this case, we have considered a submission, where tokens were written in wrong order but similarity gives 99% matching. In this query vector *q*, four tokens were written without proper sequence. The vector shown in followings: [`&n);`, `++i)`, `break;`, `else`, `flag=0;`, `flag=1;`, `for(i=2;`, `i,`, `i>=n/2;`, `if(flag==0)`, `if(n%i==0)`, `n,`, `return(0);`, `scanf("%d",`].

The cosine similarity has return 99% matching for the TF-IDF values of above token vector and fails to detect token sequence error. This is because sequence of terms never considered in TF-IDF algorithm. In this case, model also fails to contribute in program evaluation.

Case 5: It is much tedious job to develop a model with all possible solutions. In this case, we have considered a submission that was correct but due to missing in corpus, submission program was 52% matched with solution set and poorly graded. To overcome this leakage, we have added the submitted program with our solution corpus to evaluate submissions more accurately in further.

Case 4	Case 5
| ```
#include<stdio.h>
int main() {
int n, i, flag;
printf("Enter a positive integer");
scanf("%d", &n);
for (i=2; i++; i<=n/2) {
if(flag==0) {
flag=1;
break; } }
if(n%i==0)
printf("%d is a prime number", n);
else
printf("%d is not a prime number", n);
return(0); }
``` | ```
#include <stdio.h>
void main() {
int i,n,flag;
scanf("%d",&n);
for(i=2;i<=n/2;i++) {
flag=n%i;
if(flag==0) {
printf("not prime number");
goto end; } }
printf("prime number");
end:
getch(); }
``` |

4   Evaluation Methodology

We have developed three individual models having thirty solutions of each for auto-matic evaluation of prime number, binary search, and bubble sort program, respec-tively. As each model has developed with 30 solutions, hence, a new submission was compared for 30 times. Figure 4 shows the cosine similarity rate of a submitted pro-gram with each solution program. The maximum similarity was considered to grade submission and to measure model performance. We have analyzed 57 submitted pro-grams to estimate the relationship between the number of error tokens and cosine similarity. The similarity and error token relationship is not fair at all. Sometimes, wrong token leads variation on similarity rate due to tokens' rarity. Nevertheless, we were able to draw an approximate relation between them. We used linear regression model for estimation. Figure 5 shows the error rate and its affection on cosine simi-larity. Figure 6 shows cosine similarity range from 0.99 to 0.95 and defines a simple error. This happens due to initialization error or common token missing. Between 0.95 and 0.85 ranges, submission has at most two wrong tokens. We faced trouble to characterize cosine similarity from 0.84 to 0.65 range. Under this range, the number of error token varies. The possible error tokens were two to five under this range. A submission having 0.5 or lower similarity range is often considered as irrelevant or poorly correct.

5   Model Performance

TF-IDF algorithm basically performs well in information retrieval and document ranking problems. The principle of this algorithm is that weight of a term increased based on its rarity and terms is randomly compared with document corpus. There has no checking about sequence of terms during evaluation. Consequently, sometimes at second or third rank, retrieved document seems less important. So, except random and irrelevant token detection, tf-idf has excellent performance on automatic program

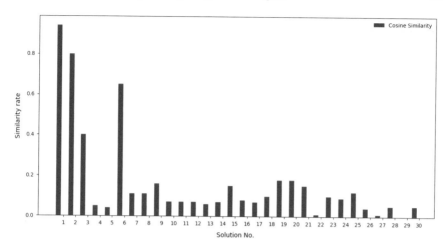

Fig. 4 Cosine similarity between submission and solution set

Fig. 5 Effect of error tokens on cosine similarity experiment

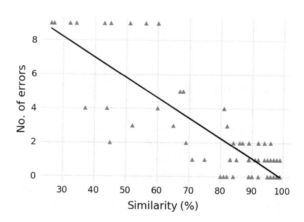

Fig. 6 Relationship between error token and similarity range(approximate)

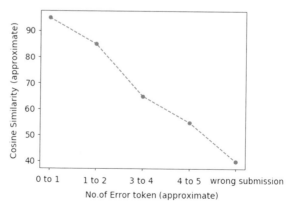

Fig. 7 Model Performance

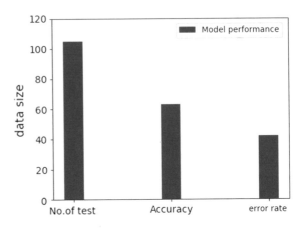

evaluation. We used 105 submitted programs to measure model performance. Among them, 63% programs were accurately evaluated like expert graders and 33% programs were misgraded that has shown in Fig. 7.

6 Conclusion and Future Work

In this paper, we have proposed a model for automatic evaluation of programming assignment using TF-IDF and cosine similarity algorithm. Without some exceptional cases, model has good performance. The performance can be increased by adding some extensions with this model. We working to improve the model performance. For improvement, we need to adapt control flow graph isomorphism and token edit distance methods. Experiments show that proposed methodologies can overcome the penalty cases of TF-IDF-based model. Now, we have been checking the vulnerability of the model by other possible cases. We believe, through adaptation of our hypothesis, this model would be able to give a higher level of performance as a replica of human expert grader.

Acknowledgements The research was done as part of MSc. Engg. in ICT in the Institute of Information and Communication Technology under a research grant by Bangladesh University of Engineering and Technology (BUET), Dhaka.

References

1. Zougari S, Tanana M, Lyhyaoui A (2016) Towards an automatic assessment system in introductory and programming courses. In: International conference on electrical information technologies, IEEE, pp 496–499

2. Zougari S, Tanana M, Lyhyaoui A (2016) Hybrid assessment method for programming assignments. In: Fourth IEEE international colloquium on information science and technology, IEEE

3. Pape S, Flake J, Beckmann A (2016) STAGE a software tool for automatic grading of testing exercise. In: Thirty eighth IEEE international conference on software engineering companion, IEEE, pp 491–500

4. Pieterse V (2013) Automated assessment of programming assignments. In: Third computer science education research conference, ACM, pp 45–56

5. Kanmani S, Radhakrishnan P (2011) A simple Journal methodology to grade c program automatically. In: International of advances in embedd system research, IJAESR, pp 73–90

6. Srikant S, Aggarwal V (2014) A system to grade computer programming skills using machine learning. In: Twentieth international conference on knowledge discovery and data mining, ACM, pp 1887–1896

7. Dadic T, Glavinic V, Rosic M (2014) Automatic evaluation of students programs. In: Conference on innovation &technology in computer science education, ACM, pp 328–328

8. Huang C-J, Chen C-H, Luo Y-C, Chen H-X, Chuang Y-T (2008) Developing an intelligent diagnosis and Assessment tool for introductory programming. J Educ Technol Soc 11(4):139–157

9. Sharma K, Banerjee K, Mandal, C (2014) A scheme for automated evaluation of programming assignments using FSMD based equivalence checking. In: Sixth IBM collaborative academia research exchange conference on I-CARE, ACM, pp 1–4

Pre-processing Techniques for Detection of Blurred Images

Leena Mary Francis and N. Sreenath

Abstract Blur detection and estimation have progressively became an imminent arena of computer vision. Along with heightening usage of mobiles and photographs, detecting the blur is purposed over to enhance or to remove the images. PrE-processing Techniques for DEtection of Blurred Images(PET-DEBI) was framed to detect the blurred and undistorted images. The frailty of Laplacian has been overcome by Gaussian filter to remove the noise of the image; then, the variance of Laplacian is calculated over the images. Through analysing the variance of the images, appropriate threshold is circumscribed and further used as limitation to define blurred and unblurred images. PET-DEBI was implemented and experimented yielding encouraging results with accuracy of 87.57%, precision of 88.88%, recall of 86.96% and F-measure of 87.91%.

Keywords Blur detection · Blur estimation · Gaussian Filter · Laplacian function · Threshold fixing

1 Introduction

Blur detection and estimation assist the field of computer vision with auto-focussing technique and quality assessment of the image. The incremented availability of good mobiles at affordable rate has benefited the people to capture increased number of digital images and to store them [6]. This has unlocked the field of blur detection, blur estimation and de-blurring of the images. Blur is defined as the phase of image where its content turns out to be difficult to read and understand. The image may become blur owing to limited contrast, untimely exposure, improper lighting environment

L. M. Francis (✉) · N. Sreenath
Department of Computer Science and Engineering,
Pondicherry Engineering College, Pondicherry, India
e-mail: leena.k@pec.edu

N. Sreenath
e-mail: nsreenath@pec.edu

© Springer Nature Singapore Pte Ltd. 2019
N. Chaki et al. (eds.), *Proceedings of International Conference on Computational Intelligence and Data Engineering*, Lecture Notes on Data Engineering and Communications Technologies 28, https://doi.org/10.1007/978-981-13-6459-4_7

Fig. 1 Stages of blur detection framework

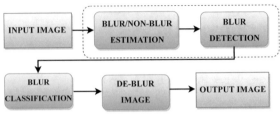

Fig. 2 Architecture of PET-DEBI

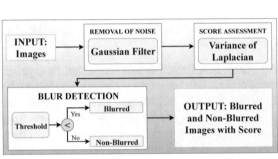

and indecorous device handling. The commoners have a trend of capturing hundreds of images daily, and it will be highly useful if a technique is skilled with automatic detection of blurred images and its deletion which will avoid the time of glancing through all the images and deleting it one by one [3]. At the same time, it will succour in increased space for the storage of images. Multitudinous accusations [10] arise from the fields like remote sensing, medical imaging, microscopy, visually impaired users and astronomy where blurred images bring about predicament in the diagnostics [4] (Fig. 1).

Figure 2 reveals the stages of the blur detection framework. The proposed work concentrates only on the stages which comprise blur/non-blur estimation and blur detection which is enclosed in red-dotted lines in Fig. 2. The proposed work, PrE-processing Techniques for DEtection of Blurred Images(PET-DEBI), was aimed as the pre-processing step in the text recognition from natural scene images, wherein there is the need to remove the blurred images in the first stage before processing into the phases of segmentation and recognition of text.

The proposed work contributes:

- To remove the noise of the image which may hinder the process of blur detection.
- To identify the optimal threshold value which forms the basis of the blur detection.
- To classify the images into blurred (B) and not blurred (NB) images based on the identified threshold value.

Section 2 lists some of the works associated in the field of blur detection, followed by the details of the proposed work in Sect. 3. Section 4 contains the experimentation and result analysis; finally, the conclusion and future enhancements are furnished in Sect. 5.

2 Related Works

Bolan Su et al. state various kinds of blur, notably motion blur, defocus blur and blend blur [7]. At the time of exposure, if there exists any relative motion between the camera and the object, then it is defined as motion blur. If the object is out of focus at the time of exposure, then its called as defocus blur. At the time of exposure, if there exists motion between the camera and object as well as the object is out of focus, then those types of blurs are named as blend blur. The stage blur detection detects the presence of blur in the image, blur estimation detects the measure of blur in the image, blur classification identifies the type of blur, and finally, the de-blur removes the blur from the image and gives the refined form of the image. PET-DEBI considers all the blur types as a single domain and detects the image if blurred stating the measure of blurriness in the image.

Dong Yang et al. [11] had made an attempt to identify the blurred regions of the image using total variation approach and had estimated the blur kernel conclusively restored the blurred region, he had simulated the experiment and had produced the results. Blur detection task is performed by Rui Huang et al. through convolutional neural networks (CNNs) [1]. The drawback with this model is that the data hungry CNN model requires a vast amount of data for training purpose, which becomes a real-time conundrum. Another research [8] performs a binary classification of blur detection into blurred and non-blurred images using Haar wavelet transform; they have used edge detection technique to identify the blurriness in the image. They also claim that their approach is 50% faster than their previous approaches.

Till Sieberth et al. proposed saturation image edge difference standard deviation (SIEDS) to detect and eliminate blurred images in order to assist the unmanned aerial vehicles (UAVs). Like human, the proposed method compares the images with other images and surmises a conclusion for the utilization of the image [5]. Van Cuong et al. proposed a fuzzy model to accomplish blur estimation on document images. The blur score is determined by the average of all the pixels in the blurred region and had evaluated on two real-time databases [2]. Through the current state of the arts, the importance of blur detection in the field of computer vision has been revealed and the various noteworthy benefits in the diversified real-time solicitations.

3 Proposed Work

The proposed work takes in images as input, and as the PET-DEBI uses Laplacian operator to determine the score of blurriness, it becomes mandatory to remove the noise of the image, as Laplacian is very sensitive to noise. The noise of the image is removed by Gaussian filter and further provided to the next module, where the variance of Laplacian operator is calculated and the resulting score is used to detect the blurred images. If the calculated score is less than the measured threshold, then it is blurred, otherwise non-blurred. Finally, output is the images that are classified into blurred and non-blurred with calculated score, which acts as aid to eliminate the blurred images.

3.1 Gaussian Filter

Gaussian filter exerts Gaussian function to remove the noise of the image. Laplacian operator is known for its sensitivity to noise which will hinder its performance, so it becomes compulsory to eliminate the noise. The Gaussian function produces a filter that can be applied to each pixel in the image. It is calculated for two dimensions as follows:

$$GF(\alpha, \beta) = \frac{1}{(2\pi\sigma^2)} e^{-\frac{(\alpha^2+\beta^2)}{2\sigma^2}} \tag{1}$$

where α is the measure between the origin and horizontal axis, and β is the measure between origin and vertical axis. σ is the Gaussian distribution standard deviation.

3.2 Variance of Laplacian

Laplacian operator is the second-order differential operator in n-dimensional Euclidean space. It is expressed as the divergence (∇) of the gradient (∇f). The Laplacian of f is the sum of the second partial derivatives in the Cartesian coordinates x_i which is defined as follows:

$$\nabla^2 f = \sum_{i=1}^{n} \frac{\partial^2 f}{\partial x_i^2} \tag{2}$$

Then -DEBI performs the variance over the resultant. The Laplacian emphasizes the regions with rapid change in intensity. It is known fact that images with very less edges are assumed to blurred images. The high variance discloses that there are more edge-like and non-edge-like objects recording high response, whereas the low variance discloses that there is little edges recording tiny spread of responses.

3.3 Threshold Computation

If the calculated variance is below the threshold, then it is concluded as "blurry"; otherwise, its "not blurry" (undistorted). The complication is setting the correct threshold as too low threshold for blur detection can erroneously mark blurry as not blurry images and too high threshold can erroneously leave the blurry images out. So the PET-DEBI has formulated out to find the proper threshold. PET-DEBI ran over the 2450 images containing undistorted and blurred images. And the variance of Laplacian for each image is measured. Then, the average of the variance of the undistorted as well as blurred images is estimated to form the near accurate threshold.

$$\frac{1}{(n+m)} \left(\sum_{i=1}^{n} \nabla^2 f(\omega) + \sum_{j=1}^{m} \nabla^2 f(\psi) \right) \tag{3}$$

where n and m mention the number of undistorted and blurred images, respectively, ω and ψ represent the set of undistorted and blurred images, and $\nabla^2 f$ specifies the variance of Laplacian.

4 Experimentation and Result Analysis

The PET-DEBI was programmed in Python, in Ubuntu distribution of Linux operating system.

CERTH dataset is functioned to perform image quality assessment, which contains 2480 digital images containing 1249 undistorted images and 1231 natural and artificial blurred images. The threshold is fixed over the training dataset and evaluated over the testing set of CERTH dataset. Along with the CERTH dataset, PET-DEBI is tested over the collection of random 2200 sample images taken from the Google.

Table 1 and Fig. 6 list out the comparison of blur detection of PET-DEBI with other such works and have given promising results with accuracy of 87.57%, precision of 88.88%, recall of 86.96% and F-measure of 87.91%. It was able to detect blurred images in five to seven microseconds.

Table 1 Comparison of blur detection methods over CERTH

Method	Accuracy(%)	Precision(%)	Recall(%)	F-Measure(%)
Giang Sontran et al. [8]	75.34	77.27	74.56	75.89
Seyfollah Soleimani et al. [6]	62.86	70.01	66.66	68.29
Siddartha Pendyala et al. [3]	81.14	88.57	71.26	78.98
Bryan M Williams et al. [9]	84.21	85.36	79.55	82.35
PET-DEBI	87.57	88.88	86.96	87.91

Fig. 3 CERTH dataset—blurred estimation score

Fig. 4 CERTH dataset—not blurred estimation score

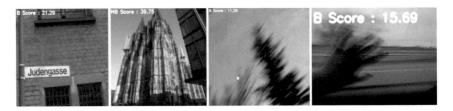

Fig. 5 CERTH dataset—incorrect estimations and random images from Google—Blurred and Non-blurred estimation score

Figure 3 is the samples of estimation of blurred images, and Fig. 4 is the samples of the estimation of clear images. In Fig. 5, the first two images show the incorrect estimations of the PET-DEBI and last two images show random estimation of the proposed work over Google images.

5 Conclusion and Future Enhancements

PET-DEBI is contrived to detect the blurred images. In order to overcome the sensitivity of Laplacian towards noise, Gaussian filter was employed over the images to remove the noise, and the variance of Laplacian was calculated over the images. If the variance is lesser than the computed threshold, then it is termed as blurred images; otherwise, they are undistorted images. PET-DEBI produced better results compared to its other works with lesser amount of time for detection. Further, the work can be extended to de-blur the images and enhance it (Fig. 6).

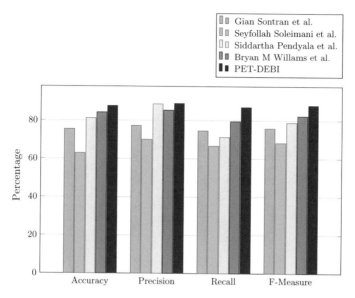

Fig. 6 Comparison over CERTH

Acknowledgements The research is funded by University Grants Commission as part of their programme called as Maulana Azad National Fellowship.

References

1. Huang R, Feng W, Fan M, Wan L, Sun J (2018) Multiscale blur detection by learning discriminative deep features. Neurocomputing 285:154–166
2. Kieu VC, Cloppet F, Vincent N (2017) Adaptive fuzzy model for blur estimation on document images. Pattern Recogn Lett 86:42–48
3. Pendyala S, Ramesha P, Bns AV, Arora D (2015) Blur detection and fast blind image deblurring. In: India conference (INDICON), 2015 Annual IEEE. pp 1–4
4. Rooms F, Pizurica A, Philips W (2002) Estimating image blur in the wavelet domain. In: IEEE international conference on acoustics speech and signal processing. IEEE; 1999 vol 4, pp 4190–4190
5. Sieberth T, Wackrow R, Chandler JH (2016) Automatic detection of blurred images in uav image sets. ISPRS J Photogrammetry Remote Sens 122:1–16
6. Soleimani S, Rooms F, Philips W (2013) Efficient blur estimation using multi-scale quadrature filters. Signal Process 93(7):1988–2002
7. Su B, Lu S, Tan CL (2011) Blurred image region detection and classification. In: Proceedings of the 19th ACM international conference on multimedia. ACM pp 1397–1400
8. Tran GS, Nghiem TP, Doan NQ, Drogoul A, Mai LC (2016) Fast parallel blur detection of digital images. In: IEEE RIVF international conference on computing & communication technologies, research, innovation, and vision for the future (RIVF), IEEE 2016 pp 147–152
9. Williams BM, Al-Bander B, Pratt H, Lawman S, Zhao Y, Zheng Y, Shen Y (2017) Fast blur detection and parametric deconvolution of retinal fundus images. In: Fetal, infant and ophthalmic medical image analysis, Springer, pp 194–201

10. Wu S, Lin W, Xie S, Lu Z, Ong EP, Yao S (2009) Blind blur assessment for vision-based applications. J Vis Commun Image Represent 20(4):231–241
11. Yang D, Qin S (2015) Restoration of degraded image with partial blurred regions based on blur detection and classification. In: 2015 IEEE international conference on mechatronics and automation (ICMA) IEEE, pp 2414–2419

Cloud Separation of a NOAA Multispectral Cyclone Image Using ICA Transformation

T. Venkatakrishnamoorthy and G. Umamaheswara Reddy

Abstract Cumulonimbus cloud detection in cyclone image is very important for forecasting the rains, weather conditions, and the direction of cyclone in remote sensing applications. Many image processing techniques are used for identification of the types of clouds using single band, multiple bands with clustering, multiple thresholding techniques and details of texture features, color values and sensor parameters of IR and VIS bands. These methods are not given sophisticated results to retrieving the object details. This proposed technique using independent component analysis for segmentation, its give satisfaction values for cloud types. The overall band details are appearing into a limited number of channels and extract the multiple objects with spatial and spectral features of cloud regions.

Keywords Independent component analysis · Albedo reflection · Texture feature · Cloud detection · IR and VIS bands

1 Introduction of Remote Sensing and Cloud Applications

Remote sensing applications are very helpful in real-life application for forecasting and prediction of weather, rainfall and atmospheric conditions [1]. It acquires huge amounts of data from the atmosphere and the surface of the earth; these data transmitted to ground stations inform raw data or image scenes. The study of atmosphere is very important for the identification of changes in temperature in day-to-day life, which is possible with polar orbital satellites which contain low-spatial resolution sensors, and it gives the spectral bands, ranging from the low frequency to high frequency ranges (visible to thermal infrared) [2].

Earth's atmosphere is an essential component in remote sensing applications; it is one of the nuisances for earth's surface because it is very difficult to study

T. Venkatakrishnamoorthy (✉) · G. Umamaheswara Reddy
Department of Electronics and Communication Engineering, Sri Venkateswara
University College of Engineering, Tirupathi 517501, India
e-mail: murthysvu407@gmail.com

© Springer Nature Singapore Pte Ltd. 2019
N. Chaki et al. (eds.), *Proceedings of International Conference on Computational Intelligence and Data Engineering*, Lecture Notes on Data Engineering and Communications Technologies 28, https://doi.org/10.1007/978-981-13-6459-4_8

earth object with changing temperature values. Cloud is also one of the objects for inflecting the earth atmospheric values. Polar orbiting remote sensing satellites are used in inter-surface parameters applications like land surface temperature (LST), sea surface temperature (SST), normalized vegetation index (NDVI), etc.

National Oceanic and Atmospheric Administration (NOAA) satellites are near circular polar orbit, it carries the Advanced Very High Resolution Radiometer on board an instrument, contains Visible, Near Infrared, Mid-wave infrared and Thermal infrared bands used for different type of applications with ground resolution 1.1 km with wide swath. These images are applied for environmental research. A single band appears as gray level image, not analyze the spectral objects in that area, so use multiple bands for getting maximum object classification values. These multispectral images are downloaded from high-resolution picture transmission (HRPT) receiving station. The total swath area can be categorized into cloud-contaminated and cloud-free objects; some objects on earth are covered by cloud areas, which is at certain height from earth objects. Due to reflection from the cloud surface, only sensors cannot capture the earthly object for cloud-contaminated areas. The IR

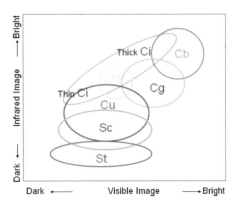

Fig. 1 Brightness of each type of clouds in VIS and IR

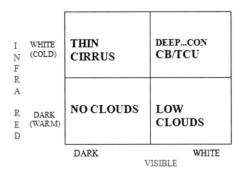

Fig. 2 Cloud matrix

channels in NOAA image illustrate the cloud categories—cirrus, cumulonimbus and nimbostratus with the help of brightness temperature. The combination of visible and IR bands discriminates the rain and no rain conditions. Temporal consistency is essential for identifying the climate conditions. Due to inconsistency between day and night, visible bands are not given priority for atmospheric applications.

Clouds are one of the main climatic objects and information about cloud area are very important, because of most atmospheric research centers, they are required cloud free fields, the clouds and its shadows are influencing the surface temperature and NDVI values of the earth objects. There are many microphysical cloud properties which are available for analyzed cloud field radiances such as cloud top height, temperature, water content, cloud fraction and cloud radiative effects. Included with this, the cloud classification is very important for identification of types of clouds from multispectral imagery for forecasting weather conditions and flooding in costal and mediterranean region (Figs. 1 and 2).

In this work, separate the cumulonimbus cloud from cyclone image, cyclone image contains very thick density clouds surrounded by cyclone eye [3]. The image processing technique is applied to all the spectral bands of NOAA image instead of individual image for getting accurate results. The study of all the individual band is very difficult, using this proposed reduction and enhancement technique study all the band details accurately with a limited number of bands [4, 5].

2 Methodology

Statistical and spectral information is needed for classifying the cloud information in this research. The land surface, sea surface and vegetation conditions are not given satisfied value by cloud-contaminated images, for identifying these clouds with sensor parameter values. Using standard traditional methods, objects are identified by the band type and sensor parameters of every channel. The main disadvantage of using this method is there is no information about cloud shape, and it is also inconvenient to retrieving the image. L-band level-2 products are widely used for identification of cloud types using AVHRR such as cloud top pressure, height, temperature and density. IAPP technique ATOVS-based cloud detection for the AVHRR cloud detection is described using temperature channels split window method. Another methods are FY-2C, FY-2B operational cloud detection method, FY2C also called Updated night cloud detection, it depends on time of capturing the data and fail in case of strong surface temperature inversions. Yamanouchi and Kawaguchi proposed the use of brightness temperature difference between thermal channels during the sunlit season. Using the multi-resolution segmentation, hierarchical classification and bi-spectral technique were performed using the multi-layers.

In previous cases, cumulus cloud mask (CCM) algorithm isolates only cumulus clouds from geostationary environment satellite (GOES); this stream is used within SATCAST to form satellite infrared (IR) cloud top temperature (TB). Mostly thermal channels are preferred for classification of clouds and atmospheric applications, thresholding, multiple thresholding techniques are applied for identification of clouds, but the tuning of threshold is difficult. Clustering techniques are mostly preferred for classifications of objects; after converting image into digital number, the pixels are not representing the exact object due to missing the some bands. Principle component analysis is also one of the best techniques for dimensional reduction, but limitation of compression of image and directionality does not giving sophisticated results. A tropical cyclone low-pressure area develops tropical waters. This low-pressure area contains energy from the warm sea surface waters. From input multispectral image, the NOAA satellite image shows a cloud pattern surrounding a cyclone. It is formed by cumulonimbus clouds, and its direction is identified for forecasting the cyclone direction. So separations of cumulonimbus clouds are essential for accurate results; mixing of cloud objects is available in that area. Visible and infrared channels are used for classification approach. Using this method, user requires the knowledge of channels and its application. The main object is mixed with cloud shade, volcanic ash and various types of clouds. These are identified by thermal channels.

i. **Principal component analysis**:
   The principal component analysis (PCA) is a statistical method for an orthogonal transformation; it converts the observations of a set of correlated variables into linearly uncorrelated variables; these variables are called principle component analysis. This technique is very suitable for dimensionality reduction and compression operations. But due to limitation of directional and non-Gaussian applications, PC1, PC2 and PC3 components are orthogonal to each other. The maximum information is available in first three components, but independent component analysis gives more satisfactory results compared to PCA [6].

ii. **HOS-based Independent Component Analysis**:
   Lower-order statistics are very useful in image comparison operations; these are very familiar up to second-order statistics. The skew and kurtosis values are third- and fourth-order statistical parameters. Compared to second-order statistics, these are given more accurate values for large object and density objects also. Instead of using a uniform distribution using this technique, the skewed values appear as main objects. For atmospheric research, the clouds are important objects; these are appearing in visible and infrared bands also. So these types of objects are enhanced dense cloud areas using the HOS technique in dimensional reduction techniques.

iii. **Clustering Technique**:
   Clustering can be done by grouping the data which are homogeneity feature, to classify the object from a given set of points; these clustering points have similar features, while data points in different groups should have highly dissimilar features. Mostly supervised or unsupervised classification techniques are used for classify the image objects. The threshold and multi-threshold techniques are used in identify the cloud contaminated and cloud free areas from mixing pixels [7]. Problems of this type are avoided with k-means clustering supervised or unsupervised algorithms. In k-means clustering algorithm, the each data will

classify the distance between two clustering groups. Recomputed the center of groups by mean of all the objects in the same group in clustering process. Same procedure repeated randomly initializes the group centers. Compared to other algorithms, k-means algorithm is fast and contains few computations and linear complexity [8].

3 Proposed Method

This paper collected NOAA-AVHRR satellite image of Cyclone Ockhi nearing its polka dot intensity in the Arabian Sea, West of India on 2 December from L-Band Receiving Station, Center of Excellence from Sri Venkateswara University, Department of Electronics and Communication.

In this technique maximum amount of information will be loss due to quantization operation. Compare this technique half of the information will be less due to quantization operation [9], so the each pixel information is very important in case of satellite images, this type of compression is not suitable in multispectral images, avoiding this problem using independent component analysis band reduction technique [10].

In cyclone image cumulonimbus clouds and lower level clouds, in the original signal visually appeared as same clouds, using image processing technique the objects selected object separated from mixed pixels. The separation of cloud-contaminated pixels and cloud-free areas is easy, but separation of types of clouds in cloud-contaminated mixing pixels is difficult. In visible channel the clouds are identified by texture features like shape, size and density of the objects. But these based on these feature not exact identify the cloud objects. This paper focused on applying image processing techniques for total multispectral image without user knowledge of all channels for detection of objects using unsupervised clustering approach [11]. The unnecessary bands give the redundant information in many channels; user requires the knowledge of data and channel frequency range application. In this technique, considering all channel information, the skewed information object is occupied as the first component in that image. In this proposed technique apply clustering for independent component analysis transformation images, using this technique first reduce dimensionality and redundant information into limited bands, in this technique the used higher order statistics skew and kurtosis operations, these are very suitable for non-Gaussian applications [12], using in this operation in multispectral image which information available in all the bands that information appear in first channel, another information in second channel likewise. So the total information is maximum in limited bands instead of all channels. After applying k-means clustering algorithm to separating homogeneity of pixels formed as same class, these same digital number objects appeared as one object. These objects are compared with albedo reflectance, surface temperature and NDVI values [13]. Generally for water objects NDVI index values have very less, based on this the cloud object values are represents low NDVI values compare with land areas. The temperature varies with respect to the density of the cloud; the dense cloud objects have higher reflectance and lower temperature values.

4 Results

Figure 3 represents the NOAA-AVHRR Ockhi Cyclone multispectral image, which is downloaded from HRPT receiver. Figure 4 shows the limited band signal which is used higher-order statistics independent component analysis for object identification [14]. Compared to multispectral input image, ICA image is more enhanced for non-Gaussian information and dimensional reduction data; ICA image contains low- and high-frequency components and enhanced spatial and spectral with limited band information.

Figure 5 gives the information about channel difference using spilt-window method. After applying threshold method for getting mixing cloud objects [9, 15], the final cyclone cumulonimbus clouds results give the equivalent result of Visible and IR resultant values. Table 1 shows the sensor values of input image objects; the NDVI and temperature values are very low; the surface reflectance values are very high compared with other clouds in selected areas. The NDVI, surface temperature and reflectance values are used for reference for comparison with proposed image (Figs. 6, 7 and 8).

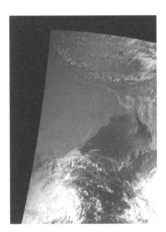

Fig. 3 Multispectral image

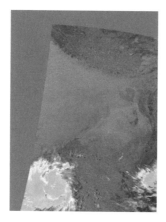

Fig. 4 Dimensional reduction technique

Fig. 5 Ch3-ch4 image

Table 1 Sensor values of NOAA multispectral image

S.No	Object	NDVI	Surface reflectance	Brightness temperature
1	Cumulonimbus cloud	−0.555	928.352	−82.20
2	Thick clouds	−0.021	887.568	−72.15
3	Water	−0.444	365.20	22.782
4	Ground	0.14	442.914	13.398
6	Ice area	−0.013	692.191	−22.903

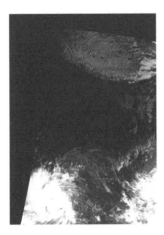

Fig. 6 Single-band image

Fig. 7 Cyclone cumulonimbus cloud image

5 Conclusion and Future Scope

Many satellite images are used for different applications in our real life, the knowing of sensor's details is very important, user requires a lot of knowledge about all the bands, and the sensors lose some information due to lack of acquisition and atmospheric information. That time sensor information will be loss and difficult to read multiple band details. The image processing techniques are more useful for separating the objects in multispectral images with limited bands. Implemented proposed technique without knowledge of band details directly apply higher order statistics based independent component analysis and k-means clustering algorithm apply for satellite images to segmented the each object with homogeneity criteria.

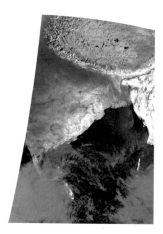

Fig. 8 NDVI image

This method is not only for satellite images, but also for medical images and time series and signal processing applications.

Acknowledgements The Authors are grateful to the Technical Education Quality Improvement Programme, Phase II, Center of Excellence [TEQIP 1.2.1 (CoE)], Sri Venkateswara University College of Engineering, for providing L-Band Receiver NOAA multispectral images and training equipment. Thanks for Vignan's foundation for science, technology and Research, Vadlamudi for providing this grateful opportunity in international conference (International Conference on Computational Intelligence & Data Engineering, ICCIDE-2018).

References

1. Liu D, Xia F (2010) Object-based classification: advantages and limitations. Remote Sens Lett 1(4):187–194
2. Benz U, Hofmann P, Willhauck G, Lingenfelder I, Heynen M (2004) Multi-resolution, object-oriented fuzzy analysis of remote sensing data for GIS-ready information. ISPRS J Photogram Remote Sens 58:239–258
3. Jaiswal N, Kishtawal CM (2011) Automatic determination of center of tropical cyclone in satellite-generated IR images. IEEE Geosci Remote Sens Lett 8:460–463
4. Kärner O (2000) A multi-dimensional. Histogram technique for cloud classification. Int J Remote Sens 21:2463–2478
5. Gillespie AR, Kahle AB, Walker RE (1987) Color enhancement of highly correlated images. II. Channel ratio and,,chromaticity transformation techniques. Remote Sens Environ 22(3):343–365
6. Gonzalez-Audicana M, Saleta LJ, Catalan RG, Garcia R (2004) Fusion of multispectral and panchromatic images using improved IHS and PCA mergers based on wavelet decomposition. IEEE Trans Geosci Remote Sens 42:1291–1299
7. Li Q, Lu W, Yang J (2011) A hybrid thresholding algorithm for cloud detection on ground-based color images. J Atmos Ocean Technol 28:1286–1296

8. Chen Pei-yu, Srinivasan Raghavan, Fedosejevs Gunar (2003) An automated cloud detection method for daily NOAA 16 advanced very high resolution radiometer data over Texas and Mexico. J Geophys Res 108(D23):4742
9. Miano J (2000) Compressed image file formats: JPEG, PNG, GIF, XBM, BMP. In: Miano J (ed) Compressed image file formats: JPEG, PNG, GIF, XBM, BMP, 2nd edn. Association for Computing Machinery, p 23
10. Hoyer Patrik O, Hyvärinen Aapo (2000) Independent component analysis applied to feature extraction from colour and stereo images. Comput Neural Syst 11:191–210
11. Hughes MJ, Hayes DJ (2014) Automated detection of cloud and cloud shadow in single-date landsat imagery using neural networks and spatial post-processing. Remote Sens 6:4907–4926
12. Venkata Krishnamoorthy T, Umamaheswara Reddy G (2017) Noise detection using higher order statistical method for satellite images. Int J Electron Eng Res 9(1):29–36
13. Visa A, Valkealahti K, Iivarinen J, Simula O (1994) Experiences from operational cloud classifier based on self-organizing map. In: Application of artificial neural networks, SPIE 2243, pp 484–495
14. Venkata Krishnamoorthy T, Umamaheswara Reddy G (2018) Fusion enhancement of multispectral satellite image by using higher order statistics. Asian J Sci Res 11:162–168
15. Inoue T, Ackerman SA (2002) Radiative effects of various cloud types as classified by the split window technique over the eastern sub-tropical pacific derived from collocated ERBE and AVHRR data. J Meteorol Soc Jpn 80(6):1383–1394

T. Venkatakrishnamoorthy is Research Scholar at Department of Electronics and Communication Engineering, Sri Venkateswara University, Tirupati. He obtained his B. Tech in electronics and communication engineering from JNT University, Hyderabad, and M. Tech in communication and signal processing from Sri Krishnadevaraya University, Anantapur. He has teaching and industrial experience of 6 years. His areas of interest are image and signal processing. He is Lifetime Member of ISTE.

Prof. G. Umamaheswara Reddy obtained his B. Tech in electronics and communication engineering, M. Tech in instrumentation and control systems, and Ph. D. from Sri Venkateswara University, Tirupati. He is Member of ISTE, IE and BMSI. Currently, he is Professor at Department of Electronics and Communication Engineering, Sri Venkateswara University, Tirupati, Andhra Pradesh. He has a teaching experience of more than 20 years and has 16 technical publications in national/international journals. His areas of interest include signal processing and biomedical signal processing

Automatic Edge Detection and Growth Prediction of Pleural Effusion Using Raster Scan Algorithm

C. Rameshkumar and A. Hemlathadhevi

Abstract Pleural effusion (PE) is the extra fluid with the purpose of accumulating between the two pleural layers and the fluid-stuffed gap so as to surround the lungs. The buildups such as fluid inside the pleural opening are commonly a symptom of an extra illness consisting of congestive heart failure, pneumonia, or metastatic cancers. Computed tomography (CT) chest examines experiment and is presently used to measure PE as radiographs and ultrasonic methods had been located to be much less correct in prediction. The proposed approach focuses on automating the process of detecting edges and measuring PE from CT scan images. The CT scanned images are processed initially to reduce intensity from the image by making it smooth. Then, edge detection algorithm is applied to that smooth image to identify visceral pleura (inner layer) along with parietal pleura (outer layer). The ending points of these two identified layers are detected using a high-speed raster scan algorithm. The pixels identified within these end points are detected to measure the affected area. This proposed is evaluated and uses advanced image processing techniques. Hence, it proves to be good implementations in clinical diagnostic purposes, as the processes are entirely computerized with time-effective.

Keywords Pleural effusion (PE) · Visceral pleura (interior layer) · Parietal pleura (external layer) · Computed tomography (CT)

1 Introduction

C. Rameshkumar (✉)
School of Computer Science and Engineering, Galgotias University, Greater Noida, Uttar Pradesh, India
e-mail: c.ramesh@galgotiasuniversity.edu.in

A. Hemlathadhevi
Department of Computer Science and Engineering, Meenakshi College of Engineering, Chennai, Tamil Nadu, India
e-mail: hemlathadhevi@gmail.com

© Springer Nature Singapore Pte Ltd. 2019
N. Chaki et al. (eds.), *Proceedings of International Conference on Computational Intelligence and Data Engineering*, Lecture Notes on Data Engineering and Communications Technologies 28, https://doi.org/10.1007/978-981-13-6459-4_9

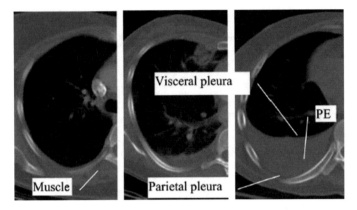

Fig. 1 Example of CT scans of PE

In the pleural effusion, the accumulating fluid that accumulates between the two pleural layers is the liquid area around the lungs. Unreasonable measure of such liquid can constrain lung prolongation amid relaxing. Our body creates little measures of pleural liquid for the pleura, chest cavities, and thin tissues around the lungs. This is an irregular inordinate liquid in pleural radiation. There are two writings: Brief pleural emanation is caused by liquid spillage into the pleural space. This is caused by circulatory strain or low blood check. Congestive heart disappointment is the most widely recognized reason for pleura emanation [1].

An unusual liquid is framed in the liquid in the lungs. Pneumonia can be caused by numerous medical issues. Most pleural emanations are not genuine, but rather some expect treatment to stay away from issues. The pleural layer is a thin film that blows the surface of the lungs and within the chest divider into the lungs. In the pleural liquid supply, liquid aggregates in the space between the pleural layers [2]. Usually, the pleural space contains only one teaspoon of aqueous liquid that allows the lungs to move smoothly under the breath in the chest cavity. Several diseases are capable of origin pleural fluid. One of the most common causes is: Excess fluid may accumulate because the body cannot cope properly with the fluid (e.g., congestive heart failure, kidney and liver disease). Fluids containing pleural fluids include inflammation, such as pneumonia, autoimmune disease, and many other conditions [3]. Rising the amount of fluid increases lung pressure and reduces pulmonary compliance leading to breathing difficulties and dyspnea in patients. Pleurocentesis relieves dyspnoea; therefore, in addition to the clinical condition, the quantitative determination of pleural effusion is of great importance in treating the patient.

PE is readily detectable and measurable in CT studies for radiologists as the liquid typically exhibits a homogeneous signal. However, this process is time-consuming and automatic segmentation faces great difficulties: The intensity of fluid from PE is almost the same as that of the surrounding tissues (see Fig. 1). This means that they have a normal growth algorithm in the region [4].

2 Literature Survey

In 2007, Segal et al. [5] the picture of the image on global gene expression in hepato-cellular carcinoma. In the CT scan of the liver cancer, 138 pictorial properties were identified and filtered using the Pearson correlation margin, with a value of 0, 9 to decrease to 32. Out of these 32 image properties, 28 were aware of gene expression. The training kit contained 28 human hepatocellular carcinomas (HCCs) with 6732 genes, and the test kit contained 19 HCCs and the 28 image traits the predicted change in expression of 6732 genes (78%). The authors conclude that genomic activity in the liver is not observed by invasive imaging.

In 2008, Al-Kadi and Watson [6] were talented to differentiate aggressive along with noninvasive malignancies (high and low down metabolism). Fifteen patients were injected with fluorodeoxyglucose, contrast agent, and at least 11 times with sequential CT images. This method provides 83.3% accuracy in identifying tumor aggressiveness.

In 2009, Samala et al. [7] looking for the optimum image features, torpid computed tomography is a thin part with a computer-aided design (CAD) system. The authors summarize the correlation of the 11 calculated image properties and 9 radar explainable characteristics of the 42 regions of interest in 38 patients. Figure 3 shows the calculation of 11 for both knots and not knots. It was found that the calculated characteristics had to be taken into account in the CAD systems and they had to be taken into account the general effect of these features.

In 2010, Ganeshan et al. [8] found that the CT output of non-little cell lung tumor connects with the digestion of glucose in the carcinogenic state. The images were first filtered using a Laplace–Gauss spatial bandwidth fine (2 pixels, 1.68 mm) and average (6 pixels, 5.04 mm) and rough (12 pixels 10.08 mm) and unfiltered. Then, there were the average gray levels, the structure of entropy, and uniformity. In 17 patients, the authors found that their rough texture is correlated with the normal value of patients absorption and fine texture is correlated with the degree. This study gives early evidence that structural features of the texture may be predictive of non-small cell lung cancer.

In 2010, Zhu et al. [9] showed that vector support can improve classification of the single pulmonary node (SPN). Using 77 reinforced biopsy CT cases, signal 67 was restored and filtered to 25 with genetic algorithm.

In 2010, Lee et al. [10] have developed a two-tiered CAD system for identifying benign and malignant lung nodes. The test was tested by cross-validation of 125 nodules. The authors ensemble method, which combines the genetic algorithm 28 selection element and the random alteer method, are better able to perform than one of the two methods to separately classify linear discriminative analysis.

In 2014, Aerts et al. [11] encouraged the idea that tomography can be seen in a noninvasive manner using a medical image. One thousand and nineteen lung, head, and neck cancer patients were studied, extracting 440 images, consisting of intensity, shape, structure, and corneal categories. Using the observed clustering, the authors found three groups of patients and found a significant correlation with primary tumor stage, general tumor stage, and histology. The proposed signature is validated by a consistency index and shows better prognosis efficacy than volume and better or comparable performance than TNM. The four signature functions are also significantly related to different gene expression profiles, indicating that they describe different biological mechanisms.

In 2015, Shen et al. [12] used a very large convolution neural network with a typical subtraction of classification of benign and malignant lumps. This method uses a lot of patches instead of a segment. The patch cube is 96, 64, and 32 pixels around the nose. These patches were placed in a very large convolution neural network. The authors found that the grading accuracy of 86.84% random forest classifiers and concludes that their method of describing descriptive features is stains.

In 2016, Lee et al. [13] carefully reviewed the radio. The authors identified the radicals of "research that can query data, perform high performance, and analyze large volumes of advanced quantitative imagery images in medical imaging," which are identified in the areas of interest characterization and build predictive models. Lung cancer is a genetic disorder that leads to regulation of cell proliferation. Genomic heterogeneity exists in tumors that produce tumorous habitats. Precise cure should be able to identify treatment responses for all tumorous habitats. The radiomics is a method of studying these habitats over time and non-invasively making it difficult to treat multiple biopsies.

In 2018, Yuan et al. [14] the proposed method uses hybrid descriptors of several major CNN statistical characteristics and uses the geometric features of s of the encoding of the Fisher vector (FV) invariance transformation element (SIFT). First, the intensity of the sample analysis and the icosahedrons intensity approach the radii of the nodes. We then use high-frequency measurements to get more information about sample views. Subsequently, based on the extracts, extracts are made to form several large CNNs that look for statistical characteristics and count the FV codes with geometric features. Finally, hybrid functions are achieved by merging MKL-based statistical and geometric features and sorting cluster types through a multi-class support machine. LIDC-IDRI and ELCAP trials have demonstrated that our techniques have accomplished promising outcomes and can be of extraordinary help in diagnosing lung malignancy in clinical practice.

In 2017, Priya et al. [15] this article introduces a computerized classification method for CT (Computed Tomography) imaging images generated with ANN-BPN. The aim of this work is to discover and classify lung diseases by transforming double complex waves and GLCM. CT images are used for segmentation of the whole lung, and the parameters are calculated from the segmented image. The parameters are calculated using the GLCM. We offer and evaluate the ANN backpropagation network that serves to classify ILD models. Parameters provide maximum accuracy for classification. After the result, fuzzy clustering is segmented from the abnormal lung of the injury part.

3 Proposed System

During CT printing, X-rays are used for recording. The patient lies in a passage while the machine turns and transmits from various beams. These photographs are later utilized by the influenced area (or cross-segment) imaging PCs. CT scanners use computers and rotary X-ray devices to create slices or lung cross sections. Unlike other techniques, CT examinations show the interior of the lung, including soft tissues, bones, and blood vessels.

CT is based on the same ideology as usual X-rays. X-rays are engrossed differently from dissimilar parts of the body. The bones swallow most X-rays, so the picture is white. The aqueous or liquid cavities are absorbed in the middle of the lungs and are black in color. As the X-ray tube as well as sensor rotates, they contain a cross section (a few millimeters broad). During the shooting, hundreds of photographs are taken, which are later used by the computer to make the final shot. During CT imaging, the X-beam tube pivots around the patient and gathers numerous pictures from various points. These pictures are stored on a computer that analyzes them to create a new image with deleted surface structures. CT imaging allows radiologists and other doctors to identify internal structures and examine their shape, size, density, and structure. This comprehensive information is able to conclude whether there is

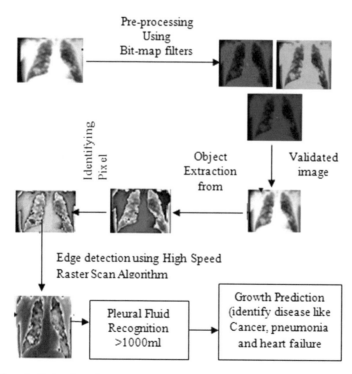

Fig. 2 Pleural effusion detection architecture

a medical predicament, the extent and the circumstances of the problem as well as additional vital details is shown in Fig. 2.

Images may still appear when there is no disorder. A CT scan that does not show anomalies still provides useful data. This information helps diagnose, focusing on unnecessary health problems. Models of contemporary topographic scanners acquire this information in seconds, sometimes once per second, depending on the examination. The CT scan provides an extremely complete picture. Material is applied to the technique, border detection of a threshold technique, which is used in the second step of the extraction process to detect the edges of the CT image. The edge detection technology is faster than threshold technology. After the perimeter detection step, volume was determined and growth was predicted by CT scan. The benefits of CT scan are

- CT filter is effortless, non-intrusive, and exact.
- Ability to show bones, delicate tissues, and veins all the while.
- CT filter gives an extremely nitty gritty picture of various kinds of tissue, and in addition lungs, bones, and veins.
- CT tests are fast and simple; they can harm inside wounds and seeping in a crisis to spare lives.
- Cost-viable imaging instrument for an extensive variety of clinical issues.
- CT is less helpless to understanding movement than MRI. CT can be performed if there is any embedded therapeutic gadget despite MRI.
- Provides ongoing pictures, making it a decent instrument for focusing on negligibly intrusive techniques.
- The CT examines determination takes out the requirement for medical procedure and careful biopsy.
- After the CT examines, there was no radiation in the patient's body.
- X-rays used in the CT scan generally do not present side effects.
- Get quick pictures.
- The wealth of clear along with existing information.
- Look at most of the body.

In the emergency space, patients can be scanned quickly so doctors can quickly assess their condition. Emergency surgery may be needed to stop internal bleeding. CT images show where the surgeons work exactly where they are. Without this data, the accomplishment of the task is significantly traded off. The danger of CT presentation is low contrasted with the advantages of very much arranged medical procedure. The CT filter gives therapeutic data about other imaging tests, for example, ultrasound, MRI, SPECT, PET, or atomic pharmaceutical. Each imaging technique has advantages and limitations. CT scan is considered as an accurate method to detect pleural effusion [16].

Proposed Implementation

Method overview is a combined process dissecting the system responsibilities that are based on the detection of pleural effusion and predicting growth rate accurately through automated process to provide proper treatment to patients [17]. The CT scanned image is an input to the system, and the pretreatment step performs the image intensity and finishing. The image is processed, and proposed approach focuses on automating the process of detecting edges and measuring PE from CT scan images. High-frequency raster scan algorithm is used for detecting the edges. Edge-extracted image is given to the next process for predicting the growth rate accurately to provide proper treatment to patients. A general overview of the method is shown in Fig. 1. The details of these processes are described in the following sections.

Preprocessing of Image Samples

Preprocessing is an initial module to load the images with the initial set of images. During this module, the scheme identifies the category of an illustration. The user will be permitted to access the processing of images through a secure login system which can be considered as an authentication. Validate the CT scan image samples, and remove intensities of the image. Filtering techniques are used to remove intensities [18]. Image is given as input to the system as shown in Fig. 3. Preprocessing step removes intensities or clusters them or takes them in different color spaces. It performs the pixel validation [19]. The filtered image is given as input for the next process.

CT scan image is given as input to the preprocessing process to remove intensity from the image. Initialize filter width, filter height, image width, and image height, and then load the image into the buffer. Each valve of pixel is multiplied with the filter value to convert the image into RGB mode. Values are indicated, and the final RGB image is produced as output as shown in Fig. 4.

Edge detection

The edges of the images will be identified and marked in this module. The algorithm is used to identify the visceral plexus (inner layer) and the parietal pleura (outer layer). Select the pixels to target. Identify the best possible positioning for each selected pixel. Raster scan algorithm is used to process the image, and optimization

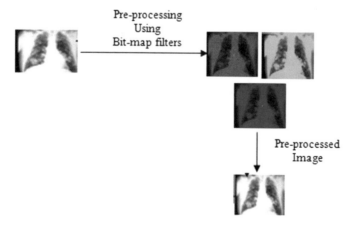

Fig. 3 Preprocessing of image samples

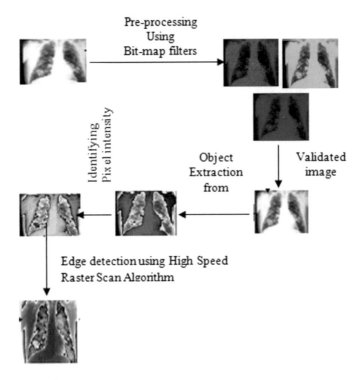

Fig. 4 Edge detection

is done only for pixels around the edges allowing a decrease of the time complexity. This module detects the edges automatically to identify pleural effusion.

Interruptions piercingly alter the strength of the pixel, which characterizes the limitations of the substance. The raster scanning algorithm processes the pixels in a row. Check each pixel valve, and check the appropriate adjacent pixel valves for edge detection as shown in Fig. 5. The principle is that not all of them become aware of the change in intensity. Effects such as refraction or poor focusing may result in objects whose boundaries are determined by gradual reduction in intensity.

The edges detection is vital in imaging procedures that utilization calculations and methods to identify and detach the coveted part or state of the picture (s). After the pretreatment was performed on a CT picture, functions can be obtained and the algorithm can be configured to detect the edges.

High-Speed Raster Scan Algorithm

Initialize the count of the number of categories by setting $C = 0$.

Consider each pixel (X,Y) in turn in a raster scan, proceeding row by row ($x = 1,2,3, \ldots,N$).

For each value of X taking $Y = 1, \ldots,n$. One of four possibilities applies to pixel (X,Y):

If (X,Y) is an edge pixel, then nothing needs to be done.

Fig. 5 Pleural fluid
recognition

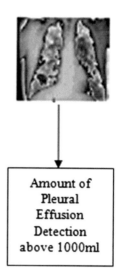

Amount of
Pleural
Effusion
Detection
above 1000ml

If both previously visited neighbors', $(X - 1, Y)$ and $(X, Y - 1)$, are edge pixels, then a new category has to be created for (X, Y).

$$C = C + 1; Q_i = C; R_j = C;$$

where the entries in $Q_1, \ldots, Q_i$ are used to keep track of which categories are equivalent, and R_j records the category label for pixel (i, j).

(3) Pleural fluid recognition: In this module, amount of pleural fluid in pleural cavity is detected automatically. In the existing system, there is no method defined to estimate the amount of pleural fluid in lungs [20, 21]. Our proposed method focus on detecting the pleural fluid level accurately to identify the type of diseases (whether cancer, pneumonia or heart failure). Pleural fluid level can be filtered from an image after detecting the edges to find volume. This process can be formulated as

$$F' = \frac{F_c - Z_1}{d - c}(F - c) + F_1$$

where Z' is the filtered image. $[F_c, F_1]$ is the available range of gray values. $[c, d]$ is the variety of concentration values in image.

Growth Prediction

The system also identifies the growth rate of the disease automatically of the particular user CT scan image. Growth level of fluid level is predicted accurately to provide proper treatment to patients. Percentage of the diseases acquired has been predicted as shown in Fig. 6.

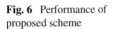

Fig. 6 Performance of proposed scheme

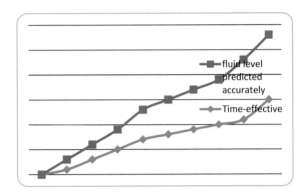

4   Conclusion

In our project, the pleural effusion detection was processed in CT scan image. Here, automated edge detection technique using an algorithm is used to identify the end points of inner and outer layers of pleura. The dermatologist grouped the pleural area in lung and further analyzed, in order to reduce the problem of pleural effusion. We have used preprocessing techniques, filtering techniques, and optimized edge detection algorithm employed in this image process to indicate the affected areas of pleural. The proposed system has more precise consequences than other existing methods.

References

1. Thawani Rajat, McLane Michael, Beig Niha, Ghose Soumya, Prasanna Prateek, Velcheti Vamsidhar, Madabhushi Anant (2018) Radiomics and radiogenomics in lung cancer: a review for the clinician. Lung Cancer 115:34–41
2. Hawkins SH (2017) Lung CT radiomics: an overview of using images as data. Ph. D. diss., University of South Florida
3. Depinho RA, Paik J, Kollipara R (2018) Compositions and methods for the identification, assessment, prevention and therapy of cancer. U.S. Patent Application 15/495,311, filed 15 Mar 2018
4. Miloseva L, Milosev V, Richter K, Peter L, Niklewski G (2017) Prediction and prevention of suicidality among patients with depressive disorders: comorbidity as a risk factor. EPMA J Suppl 8(1):1–54
5. Segal E, Sirlin CB, Ooi C et al (2007) Decoding global gene expression programs in liver cancer by noninvasive imaging. Nat Biotechnol 25:675–680
6. Al-Kadi OS, Watson D (2008) Texture analysis of aggressive and nonaggressive lung tumor CE CT images. IEEE Trans Biomed Eng 55(7):1822–1830
7. Samala R, Moreno W, You Y, Qian W (2009) A novel approach to nodule feature optimization on thin section thoracic CT. Acad Radiol 16(4):418–427
8. Ganeshan B, Abaleke S, Young RCD, Chatwin CR, Miles KA (2010) Texture analysis of nonsmall cell lung cancer on unenhanced computed tomography: initial evidence for a relationship with tumour glucose metabolism and stage. Cancer Imaging 10(1):137–143

9. Cortes C, Vapnik V (1995) Support-vector networks. Mach Learn 20(3):273–297
10. Lee MC, Boroczky L, Sungur-Stasik K et al (2010) Computer-aided diagnosis of pulmonary nodules using a two-step approach for feature selection and classifier ensemble construction. Artif Intell Med 50(1):43–53
11. Aerts HJWL, Velazquez ER, Leijenaar RTH et al (2014) Decoding tumour phenotype by noninvasive imaging using a quantitative radiomics approach. Nat Commun 5:4006
12. Shen W, Zhou M, Yang F, Yang C, Tian J (2015) Multi-scale convolutional neural networks for lung nodule classification. In: Ourselin S, Alexander DC, Westin C-F, Cardoso MJ (eds) Information processing in medical imaging: 24th international conference, IPMI 2015, Sabhal Mor Ostaig, Isle of Skye, UK, 28 June–3 July 2015, Proceedings. Springer International Publishing, Cham, pp 588–599
13. Lee G, Lee HY, Park H, et al (2016) Radiomics and its emerging role in lung cancer research, imaging biomarkers and clinical management: state of the art. Eur J Radiol (2016 Article in Press)
14. Yuan J, Liu X, Hou F, Qin H, Hao A (2018) Hybrid-feature-guided lung nodule type classification on CT images. Comput Graph 70:288–299
15. Priya CL, Gowthami D, Poonguzhali S (2017) Lung pattern classification for interstitial lung diseases using an ANN-back propagation network. In: International conference on communication and signal processing (ICCSP), 2017. IEEE, pp 1917–1922
16. Usta E, Mustafi M, Ziemer G (2009) Ultrasound estimation of volume of postoperative pleural effusion in cardiac surgery patients. Interact Cardiovasc Thoracic Surg 10:04–207
17. Harikumar R, Prabu R, Raghavan S (2013) Electrical impedance tomography (EIT) and its medical applications: a review. Int J Soft Comput Eng (IJSCE) 3(4):193–198
18. Porcel J, Vives M (2003) Etiology and pleural fluid characteristics of large and massive effusions. Chest 124:978–983
19. Rogowska J (2000) Overview and fundamentals of medical image segmentation. In: Bankman IN (ed) Handbook of medical imaging, processing and analysis. Academic, New York, NY, USA, p 6
20. Balik M, Plasil P, Waldouf P, Pazout J, Fric M, Otahal M, Pachl J (2006) Ultrasound estimation of volume of pleural fluid in mechanically ventilated patients. Intensive Care Med 32:318–321
21. Roch A, Bojan M, Michelet P, Romain F, Bregeon F, Papazian L, Auffray J-P (Jan 2005) Usefulness of ultrasonography in predicting pleural effusions >500 mL in patients receiving mechanical ventilation. Clin Invest Crit Care 127(1):224–232

Alzheimer's Ailment Prediction Using Neuroimage Based on Machine Learning Methods

Raveendra Reddy Enumula and J. Amudhavel

Abstract The unusual functionality of the brain, which distracts and causes hazardous problems in the brain, is Alzheimer's disease (AD). For decades, the level of research and identifying the disease in early stages are very less. To overcome this problem, we had introduced machine learning methods which are dedicated to the early identification of AD along with predictions of its progression in mild cognitive impairment. Alzheimer's Disease Neuroimaging Initiative (ADNI) is used for the gathering of information. A few regions of research like multi-area examination of cross-sectional and longitudinal FDG-PET images provide some reliable source which is secured at a single time point is used to improve classification results similar to those gathered using data from research quality MRI.

Keywords Neuroimage · Machine learning · Classification · Prediction

1 Introduction

The German doctor Alois Alzheimer put his endeavors on the AD and expressed as constant dementia and the adequate presence in the cerebrum of particular neuropathological structures. Generally it is identified by the symptoms such as generally memory loss, followed by further functional and cognitive decline, that patients are slowly less able to perform basic tasks [1]. In senior citizens, AD is generally a reason for dementia, crosswise over world omnipresence, that is, having an arbitrary development of the 26.6 million revealed in 2006 to more than 100 million by 2050 [2].

The AD is translated by NINCDS-ADRDA Alzheimer's Criteria [3], and the patients are ordered as having an unmistakable, likely, or conceivable AD. A conclusion of clear AD requires that neuropathological discoveries be affirmed by an

R. R. Enumula (✉) · J. Amudhavel
K L University, Green Fields, Vaddeswaram, India
e-mail: nani.naniravi@gmail.com

© Springer Nature Singapore Pte Ltd. 2019
N. Chaki et al. (eds.), *Proceedings of International Conference on Computational Intelligence and Data Engineering*, Lecture Notes on Data Engineering and Communications Technologies 28, https://doi.org/10.1007/978-981-13-6459-4_10

89

immediate examination of cerebrum tissue tests, which might be acquired either at dissection or from a mind biopsy.

The demonstrative stages are a pre-symptomatic determination, differential conclusion, the evaluation and forecast of movement. Perceptions have yielded that biochemical and neuroimaging bookmakers to have analytic and prognostic incentive for the AD, which are expressed in distributed modifications to the agreement criteria expect to join accessible these advances [4]. The AD is translated by NINCDS-ADRDA Alzheimer's Criteria; the patients are arranged as having a distinct, likely, or conceivable AD. A determination of clear AD requires that neuropathological discoveries be affirmed by an immediate investigation of mind tissue tests, which might be acquired either at post-mortem examination or from a cerebrum biopsy.

Numerous researchers are working to find biological markers (biomarkers) that indicate the presence of the disease. According to AD Neuroimaging Initiative [5], an ideal AD biomarker should be able to identify features of the pathophysiologic processes active in the AD before symptom onset. Also, it should be accurate, reliable, valid, and minimally invasive.

There is no biomarker that can analyze and predict the early stages of the AD, and observations have been made to identify and detect neuropathological processes which help in recognizing the people with the chances of having the symptoms of dementia. [6] have observed many time-dependent models of AD biomarkers which are related to elders. Accurate biomarkers which are used in AD studies are categorized as biochemistry, genetics, neurophysiology, and neuroimaging. Bookmakers have to be discovered. Biomarkers included in the review are simply those that have been proposed so far. As the researchers are increasing, there will be an advancement in the biomarkers.

As indicated by the specialists, investigation neuroimaging procedures particularize an approach to examine the basic and utilitarian changes in the mind related to the development of the AD. Usually, utilized modalities include attractive reverberation imaging (MRI), X-beam processed tomography (CT), positron discharge tomography (PET), single-photon emanation figured tomography (SPECT), and dispersion tensor imaging (DTI). The perceptions in this theory will target PET and MRI.

2   Literature Review

The research emphasize the diagnosis of AD based on imaging and machine learning techniques. [7] Authors have founded and analyzed on many different machine learning techniques and come up with different segmentation and machine learning techniques which are merely used for the diagnosis of the AD and are reviewed inclusive of thresholding, supervised and unsupervised learning, probabilistic techniques, Atlas-based approaches, and fusion of different images.

Machine learning techniques offer novel methods to estimate diagnosis and clinical outcomes at an individual level, which are also divided into supervised and unsupervised techniques. As depicted in the above Fig. 1, different supervised and

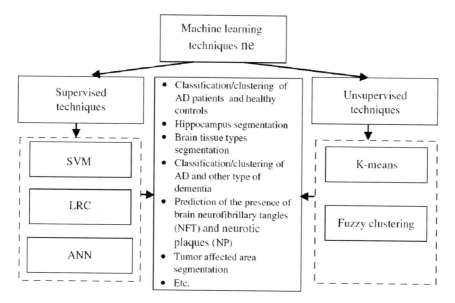

Fig. 1 Machine learning techniques used in the AD studies

unsupervised techniques are used in AD studies. Machine learning devices have been widely connected to the distinguishing proof of neurologic or neuropsychiatric issue, particularly Alzheimer's malady (AD) and its prodrome, and mellow intellectual weakness (MCI).

The automated differentiation of AD and FTD from normal subjects with the idea of selecting an optimal and best combination of Eigen brains to obtain the features that had the optimal differentiation ability. The creators have revealed a hereditary calculation-based technique to perceive an ideal blend of eigenvectors, so the resultant highlights are having the capacity of effectively part patients with suspected Alzheimer's sickness and frontal transient dementia from typical controls. Compared the approach in [8] with the standard PCA on a set of 210 clinical cases and improved the performance in separating the dementia types with an accuracy of 90.0% and a Kappa statistic of 0.849.

A nonlinear examination of complex information is a substantial approach in making free from disarray or vague about the part of the NP and NFT in the change of a procedure prompting AD [9]. Our investigation demonstrates that ANNs can precisely the location of the nearness of AD pathology. The versatile frameworks had a better precision when thought about models of customary direct insights. Both NFTs and decrepit plaques are the essential lessons of the AD. Matured plaques incorporate both diffuse plaques and NPs; generally, the greater parts of decrepit plaques are the diffuse sort.

Reference [10] proposed about the prediction, diagnosis of Alzheimer's disease (AD) and it is needed internal understanding of the disease and its progression. The scientists have moved toward the territory of the issue from various headings by endeavoring to build up (a) neurological (neurobiological and neurochemical) models, (b) diagnostic models for anatomical and useful mind pictures, (c) scientific component extraction models for electroencephalograms (EEGs), (d) order models for positive distinguishing proof of AD, and (e) neural models of memory and memory disability in AD. This article exhibits a best in the class audit of research performed in computational displaying of the AD and its markers. Mild cognitive impairment (MCI) is an intellectual issue described by memory misfortune, more than anticipated by age [11]. Another philosophy is proposed to distinguish MCI patients amid a working memory undertaking utilizing MEG signals. That new system comprises four stages: In stage 1, the total group experimental mode disintegration (EMD) is utilized to part the MEG motion into an arrangement of versatile sub-groups as indicated by its contained recurrence data. In stage 2, a nonlinear dynamic measure in light of change, entropy (PE) investigation, is utilized to dissect the subbands and observe the accomplishment client to be utilized for MCI discovery. In stage 3, an examination of fluctuation (ANOVA) is utilized to highlight determination. In stage 4, the improved probabilistic neural system (EPNN) classifier is connected to the chosen highlights to recognize MCI and solid patients. The handiness and adequacy of the proposed approach are approved utilizing the detected MEG information acquired tentatively from 18 MCI and 19 control patients.

MRI information for scholarly ordinary and Alzheimer's sickness (AD) patients from the AD Neuroimaging Initiative database was utilized as a part of this examination [12]. The survey here the evil posedness of this order issue particular measurements and test sizes and its connection with the execution of regularized strategic relapse (RLR), direct help vector machine (SVM), and straight relapse classifier (LRC). With Odin, these strategies were contrasted and their foremost part space partners.

An approach used the voxel intensities (VI) of each brain scan as the classification features. Orderly to select the subset of features worn by the classifier, all the trademarks were ranked corresponding to their Mutual Information (MI) to the class label and the highest ranking attributes were then stipulated [13]. Let $xi \in n$ signify the training patterns, $I = 1, P$ and $v \in \{1, 2 \text{ and } 3\}$ indicate the corresponding classification. This feature preference method remains used in both the favorite class ensemble and in the base classifier approach.

3 Comparative Analysis

Author	Dataset	Algorithm/method	Accuracy/result
Golrokh Mirzaei	MRI Images	SVM & UNSVM learning, Atlas-based approaches	Among the classification techniques investigated, SVM seems to provide a higher accuracy rate in the AD studies
Y. Xia	Dementia phase; symptomatic, pre-dementia phase; and asymptomatic, preclinical phase of the AD	Genetic algorithm	Achieved a classification accuracy of 90.0% and a Kappa statistic of 0.849, a very good agreement between the automated classification result and the diagnosis given by neurologists
Carlos Cabral	FDG-PET images	SVM algorithm	A peak accuracy of 66.78, 66.33 and 64.63% for ensembles of RBF-SVM, L-SVM and RF, respectively
Buscema	Neurofibrillary tangles (NFTs), neuritic plaques (NPs)	ANN	Optimization systems reached the best result with a mean accuracy equal to 90.36%
Ramon Casanova	Structural MRI data	SVM and LR C	Excellent results in terms of prediction performance comparing very well with SVM and LRC approaches
Adele	Anatomical images and functional images	ANN and LDA	Sum of the ANN algorithms already reported during the past decade. The accuracy of three algorithms ANNs—back propagation, fuzzy back propagation network and radial basis function neural network reported greater accuracy than LDA

4 Conclusion

The earliest finding of AD and MCI is fundamental to understanding, consideration, and research, and it is generally acknowledged that preventive measures assume an essential part to defer or reduce the movement of the AD. The work is displayed in the picture-based grouping of AD and MCI. The order comes about recognizing AD patients, and HC might meet on an unreasonable impediment since the analytic agreement criteria they have a precision of around 90%. Multi-district examinations of cross-sectional and longitudinal FDG-PET pictures from ADNI are performed. Data separated from FDG-PET pictures procured at a solitary time point are utilized to accomplish order comes about equivalent with those got utilizing information from a look into quality MRI.

References

1. An Atlas of Alzheimer's Disease Hardcover – October 15, (1999) by M.J. de Leon
2. Forecasting the global burden of Alzheimer's malady. Ron Brookmeyer (2007)
3. Clinical analysis of Alzheimer's malady, Guy McKhann (1984)
4. Introduction to the proposals from the National Institute on Aging-Alzheimer's affiliation work-groups of indicative rules for Alzheimer's malady. Clifford R. Jack
5. The Alzheimer's Disease Neuroimaging Initiative Susanne. G. Mueller
6. Biomarker Modeling of Alzheimer's Disease. Clifford R. Jack, Washington University, 660 S. Euclid
7. Golrokh Mirzaei (2016) Imaging and machine learning methods for the examination of Alzheimer's affliction. June 19 2016
8. Xia Y (2008) Hereditary figuring based PCA Eigenvector decision and weighting for robotized identification of dementia using FDG-PET imaging. IEEE, Designing in Medicine and Biology Society, pp 481–4815
9. Buscema (2004). Counterfeit neural frameworks and simulated life forms can expect Alzheimer pathology in singular patients just on the introduction of intellectual and useful status. NeuroInformatics 2:399–416
10. Adele (2005) Alzheimer's sickness and models Of Computation: imaging, characterization, and neural models. J Alzheimers Dis 7:255–262
11. Amezquita-Sanchez (2016) Another logic for Robotized examination of delicate subjective handicap (MCI) using magnetoencephalography (MEG). Behav Cerebrum Res 305:174–180
12. Classification of Structural MRI Images in Alzheimer's disease from the Perspective of III-Posed Problems. Ramon Casanova, Neuroimaging Initiative, October 10 (2012)
13. Classification of Alzheimer's disease from FDG-PET images utilizing Favorite Class Ensembles. Carlos Cabral 1, Neuroimaging Initiative

Formalization of SOA Design Patterns Using Model-Based Specification Technique

Ashish Kumar Dwivedi, Santanu Kumar Rath
and Srinivasa L. Chakravarthy

Abstract Due to the increase in the complexity of the current system, the need for developing solution of recurring design problems becomes more prominent. Nowadays, various analysis and design methodologies are available to resolve the complexity of the modern system. But most of them are informal and semiformal in nature. One of them is design pattern that produces reusable solution for the recurring design problem. Solution based on design pattern is specified by using UML diagrams, which are further analyzed by using formal notation, because UML is semiformal in nature. In this study, the analysis of the behavioral aspect of service-oriented architecture design pattern is presented. For the formalization of pattern notations, object modeling language, i.e., Alloy, has been used. For demonstrating the proposed approach, a case study, i.e., purchase order system, has been taken into consideration.

Keywords Alloy · Design patterns · Formal methods · SOA

1 Introduction

A good number of technologies are available to realize a service-oriented architecture (SOA) for different applications. The components of SOA are service requester, service provider, and service registry, which help in integrating with the diverse systems by providing an architectural style. The dynamic aspect of SOA is realized by

A. K. Dwivedi (✉) · S. L. Chakravarthy
Department of CSE, Gayatri Vidya Parishad College of Engineering (A),
Visakhapatnam, Andhra Pradesh 530048, India
e-mail: shil2007@gmail.com

S. L. Chakravarthy
e-mail: chakri.ls@gvpce.ac.in

S. K. Rath
Department of CSE, National Institute of Technology Rourkela Rourkela, Odisha 769008, India
e-mail: skrath@nitrkl.ac.in

© Springer Nature Singapore Pte Ltd. 2019
N. Chaki et al. (eds.), *Proceedings of International Conference on Computational Intelligence and Data Engineering*, Lecture Notes on Data Engineering and Communications Technologies 28, https://doi.org/10.1007/978-981-13-6459-4_11

Web services, which enable organizations to share information using Simple Object Access Protocol (SOAP) messages. SOA has a number of design principles that help to ensure the consistent realization of design characteristics. Design pattern is considered to develop object-oriented system with the help of design principles [1]. To develop a SOA application, various design patterns have already been proposed [2]. The available patterns have been presented by using informal and semiformal notations, which may be ambiguous in nature. The formal method is one of the mathematical techniques, which helps to verify and validate the design-level documents [3].

In the presented approach, a pattern-based method for e-business processes is proposed. A layered architecture of SOA design pattern is presented. In order to realize these e-business patterns, two GoF behavioral patterns, i.e., composite pattern and state pattern, and one security pattern, i.e., secure proxy pattern, are considered. UML class diagram is used to formalize a collection of patterns. To formalize the SOA design patterns, a formal approach, i.e., Alloy, is considered [4, 5]. A SOA-based modeling framework is proposed which promotes architect and designer to formalize SOA-based system.

2   Related Work

Tounsi et al. [6] proposed a rigorous approach to formalized message-oriented SOA. They used Event-B for formalization. Dong et al. [7] have formalized design patterns for the Web service using stereotypes notations. Kim and Carrington [8] have proposed a role-based formal modeling concept to formalize design patterns. Dwivedi and Rath [9] presented a formal approach to analyze security design patterns using Alloy. Pradhan et al. [10] have analyzed patterns using graph-based methods. Dwivedi et al. [11] have analyzed patterns using the reverse engineering method. Dong et al. [12] have formalized security patterns using Calculus of Communicating Systems (CCS). Dou et al. [13] have a proposed a metamodeling technique to formalize design patterns.

3   Proposed Work

SOA design patterns enable software architect to develop e-business applications via the reuse of component and solution elements from proven successful experiences. Pattern-based development is a collection of assets, which contain various SOA design patterns, such as business patterns, integration patterns, and composite patterns. Patterns are associated with business processes, such as purchase order, price calculation, and process scheduling. A pattern language for the e-business processes is presented in Fig. 1. Business patterns determine the interactions between business processes and data, whereas integration patterns connect business patterns

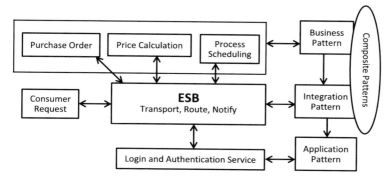

Fig. 1 Layered architecture of SOA-based patterns

when a real-world problem requires the solution of multiple business patterns. Composite patterns are generally used in orchestration and choreography of services. It combines business patterns and integration patterns.

3.1 Design Patterns for SOA-Based System

In this study, a concept of security patterns is also introduced to provide secure communication between business processes; *strategy* and *state* behavioral patterns have been considered to analyze the behavioral aspect of SOA-based application. The purchase order application has various services, which are composed using SOA-based patterns. A service is a key element in a service-oriented architecture which maps to a business function that is identified during business process analysis. Each service has an interface to enable publish, discover, and invoke processes. A service contains a number of components which perform the business function that the service represents. In this application, *purchase order* is a main component using three interfaces such as price calculation, shipping request, and process scheduling to realize respective services such as pricing service, shipping service, and scheduling service. These services are composed by using strategy design pattern which are shown in Fig. 2.

Figure 2 contains two GoF design patterns, i.e., strategy and state patterns, and one security pattern, i.e., secure proxy. In this study, *compositor* acts as a strategy for a context i.e., *composition*. State design pattern is considered to handle various states of a purchase order component for changing the properties of a component, whereas secure proxy pattern provides a secure session to perform communication process between Web services and service consumer and providers. In the context of selected patterns, the problem is the verification of purchase order component that verifies user's information. One solution is that the user's information may be verified by using security rules.

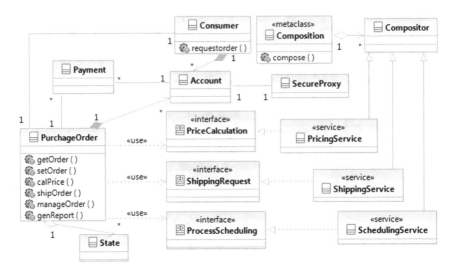

Fig. 2 Structural aspect of SOA-based patterns

3.2 Formalizing SOA Design Patterns

A pattern-based formal modeling framework is presented in Fig. 3. In the proposed framework, requirements of a system are transformed into UML-based analysis model, which can easily be transformed into Extensible Markup Language Metadata Interchange (XMI) code by using Eclipse plug-in. This XMI code is being transformed into a pattern-based design model. The pattern-based design model is nothing but a collection of UML diagrams which needs to be transformed into Alloy model. This Alloy model specifies system requirements using fact, assertion, and predicate, which are analyzed by using Alloy Analyzer (Fig. 4).

The partial Alloy model is presented to formalize SOA design patterns. The model includes fact and predicate to specify the consistency of OrderProcessing task. The predicate specifies prestate and poststate of an operation such as in predicate—prestate is *idle*, and poststate is busy. ValidateUser, RequestOrder, ProcessOr-

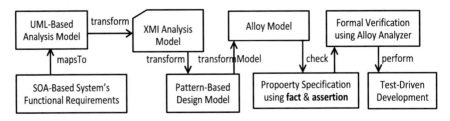

Fig. 3 SOA pattern-based formal modeling framework

sig *SOADesignPattern* {*state* : **one** *State*}
sig *Send* {*sender* : **one** *Process*, *receiver* : **one** *Process*, *cooki* : **set** *Cookie*}
{*sender* **in** *Consumer* => (**all** *ck* : *Cookie* | *ck* **in** *cooki* && *ck.server* =
        receiver && *sender* ! = *receiver*)}
pred *OrderProcessing*[*st*, *st'* : *State*, *user* : *User*, *op* : *Operation*]
{*st.state* = *Idle* && *op.operation* = (*RequestOrder* + *ValidateUser*+
        ProcessOrder) && *st'.state* = *Busy*}
fact { **all** *s1*, *s2* : *Server* | *s1* ! = *s2* && *s1.ck* ! = *s2.ck*}
assert { **all** *st*, *st'* : *State* | **all** *user* : *User* | **all** *op* : *Operation* |
        OrderProcessing[*st*, *st'*, *user*, *op*]}
run *OrderProcessing* **for** 3

Fig. 4 Alloy model for SOA design patterns

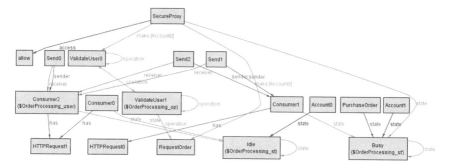

Fig. 5 Analysis result generated from Alloy analyzer

der, etc., operations are specified by using predicates. Alloy Analyzer generates the instances of a model and their relationship. The proposed model presents OrderProcessing task for problem size 3 as presented in Fig. 5.

Table 1 Performance evaluation using various SAT solvers

S. No.	# Instances	# SAT solver	Time in ms	# Test cases	# Variables
1.	6	MiniSat	521	243	68,697
2.	6	MiniSat prover	243	17	68,696
3.	6	ZChaff	491	243	68,697
4.	6	SAT4 J	600	243	68,697

3.3 Evaluation

Alloy analyzer supports a number of SAT solvers, which are shown in Table 1. These solvers execute SAT formulas. The comparison of performance between these solvers for speci c type of a system configuration is presented in Table 1.

4 Conclusion and Future Work

In the presented approach, SOA design patterns are formalized by using Alloy.

Alloy analyzer is used for the process of model checking. Nowadays, various applications are based on SOA; hence, it needs a reusable solution that is properly analyzed by using a rigorous technique. The concept of proposed formal modeling becomes helpful for similar types of system. Alloy-based modeling provides the consistency and understandability to resolve the complexity of a target system. The analysis of SOA design patterns enables users to be aware of exactly where and how to use patterns.

The proposed work may be extended to formalized Web service security and cloud patterns.

References

1. Gamma E, Helm R, Johnson R, Vlissides J (1995) Design patterns: elements of reusable object-oriented software. Addison-Wesley
2. Erl T (2009) SOA design patterns, 1st edn. Prentice Hall PTR, Upper Saddle River, NJ, USA
3. Woodcock J, Larsen PG, Bicarregui J, Fitzgerald J (2009) Formal methods: practice and experience. ACM Comput Surv (CSUR) 41:1–36
4. Jackson D (2002) Alloy: a lightweight object modeling notation. ACM Trans Softw Eng Methodol 11:256–290
5. Group SD (2010) Alloy analyzer 4. http://alloy.mit.edu/alloy4/
6. Tounsi I, Hadj Kacem M, Hadj Kacem A, Drira K (2015) A refinement-based approach for building valid soa design patterns. Int J Cloud Comput 24:78–104
7. Dong J, Yang S, Zhang K (2007) Visualizing design patterns in their applications and compositions. Softw Eng, IEEE Trans 33:433–453
8. Kim SK, Carrington D (2009) A formalism to describe design patterns based on role concepts. Formal Aspects Comput 21:397–420
9. Dwivedi AK, Rath SK (2015) Formalization of web security patterns. INFOCOMP J Comput Sci 14:14–25
10. Pradhan P, Dwivedi AK, Rath SK (2015) Detection of design pattern using graph isomorphism and normalized cross correlation. In: Contemporary computing (IC3), 2015 Eighth international conference on, IEEE, pp 208–213
11. Dwivedi AK, Tirkey A, Rath SK (2018) Software design pattern mining using classification-based techniques. Frontiers Comput Sci, 1–15

12. Dong J, Peng T, Zhao Y (2010) Automated verification of security pattern compositions. Inf Softw Technol 52:274–295
13. Dou L, Liu Q, Yang ZY (2013) A metamodeling approach for pattern specification and management. J Zhejiang Univ Sci C, Springer 14, 743–755

Earthosys—Tsunami Prediction and Warning System Using Machine Learning and IoT

Gautham Pughazhendhi, Amirthavalli Raja, Praveen Ramalingam
and Dinesh Kumar Elumalai

Abstract In this paper, the model for predicting tsunami using machine learning classification algorithm and the warning system using IoT has been proposed. The data used for training the model is based on the historical tsunami data comprising tsunami records from 2100 BC. The model has been trained based on the earthquake parameters in the dataset, as earthquakes are the main cause of death-causing tsunamis around the planet. It can classify the earthquake data as Tsunamigenic or Non-Tsunamigenic based on which the warning system is triggered, using the cloud technologies for communication. The model has been tested on the tsunami-causing earthquake records, and it shows an accuracy rate of over 95%. The model and the warning system together can automate the manual tsunami prediction techniques and alert system for the same, to save humankind in a better way than the past.

Keywords Tsunami · Earthquake · Machine learning · IoT · Cloud

1 Introduction

Tsunami is a set of destructive waves which are caused due to large, rapid disturbance on the sea surface and the most disastrous natural hazard on the planet [1]. Earthquakes are the major source for such destructive tsunamis which cause huge damage to the coastal regions around the epicenter [1]. The earthquakes that occur in oceanic and coastal regions due to the movement of tectonic plates produce faults

G. Pughazhendhi (✉) · A. Raja · P. Ramalingam · D. K. Elumalai
Velammal Engineering College, Chennai, India
e-mail: gautham.pughazhendhi@gmail.com

A. Raja
e-mail: amirthavalli.senthil@gmail.com

P. Ramalingam
e-mail: praveen.ramalingam97@gmail.com

D. K. Elumalai
e-mail: dinesh.kumar.elumalai97@gmail.com

© Springer Nature Singapore Pte Ltd. 2019
N. Chaki et al. (eds.), *Proceedings of International Conference on Computational Intelligence and Data Engineering*, Lecture Notes on Data Engineering and Communications Technologies 28, https://doi.org/10.1007/978-981-13-6459-4_12

on the ocean floor. These faults result in the vertical movement of the ocean floor which releases an enormous amount of energy from the earth's surface to the ocean [1, 2]. As a result, the tsunami is propagated as a set of waves in all directions around the epicenter. Initially, the wavelength of the tsunami waves generated is large but, as the ocean depth decreases, the wavelength of the waves also decreases with an increase in wave height [3]. An earthquake has to satisfy certain criteria to cause a tsunami. The major cause of tsunami according to International Tsunami Warning Centre is the large and shallow earthquakes having an epicenter or fault line on or closer to the ocean floor [4]. Such earthquakes are termed as Tsunamigenic, while others are termed as Non-Tsunamigenic [5]. The Earthosys model is responsible for classifying the incoming earthquake data as Tsunamigenic and Non-Tsunamigenic based on its experience from learning.

1.1 Factors of an Earthquake-Affecting Tsunami

The major characteristics of an earthquake that affect tsunami are the magnitude of the earthquake in Richter scale known as "Richter magnitude" must be large, the earthquake must have a shallow focal depth beneath the ocean floor or coastal region and the region of epicenter must be on the ocean floor or near to it [6, 7]. The magnitude of an earthquake is defined as the unit to measure the relative strength or size of an earthquake, whereas the focal depth of an earthquake is defined as the depth of the epicenter from the earth's surface, and the epicenter is a point on the earth's surface vertically above the rupture point or focus of the earthquake [8]. These were the key considerations considered while developing the model to predict tsunami.

1.2 Motivation and Contribution

The motivation behind creating Earthosys is to safeguard lives from tsunamis, the most destructive natural hazard by warning humans as soon as possible after a Tsunamigenic earthquake has been observed. This research was initiated with an optimistic intention to serve and help mankind to safeguard their lives from the tsunami. This research is augmented with the usage of modern technologies such as machine learning, IoT, and cloud to reach its full potential in an attempt of finding a solution for minimizing the damages caused by the tsunami to the human community.

Section 1 provides a brief introduction to the tsunami and the factors of an earthquake-affecting tsunami. Section 2 details about the literature survey in the field of tsunami prediction. Section 3 describes the proposed model. Section 4 describes machine learning algorithm used for building the model. Section 5 describes the warning system used in Earthosys, and in Sect. 6, the evaluation part of the model is discussed. Finally, the conclusion and the future work for this proposed model have been detailed.

2 Related Work

This section briefs about the related work achieved in predicting and warning tsunamis earlier by artificially intelligent systems. Initially, tsunamis were predicted by manual human calculation based on the seismic data retrieved from the sensors deployed all over the world.

Later with the advancement of technology, artificially intelligent fuzzy inference systems were developed for predicting tsunamis and to generate immediate alerts [9, 10]. The fuzzy inference system used for predicting tsunami was based on the fundamental concept of fuzzy logic. Fuzzy logic systems emulate a human way of thinking and reasoning to the problem under consideration. They accomplish human thinking and reasoning based on the data available for a given problem with the help of a set of rules carved by domain experts to handle the fuzzy set of data fed into the system. They rely on humans in framing the set of rules that help in producing the fuzzy output, and so, it requires a lot of human intervention in framing and editing these rules. The fuzzy logic way of approach to a real-world problem involves three important stages. The first stage involves the fuzzification process which converts the available crisp data into fuzzy data or membership functions which can be processed by the system [9]. This is followed by the fuzzy inference process which combines the rules framed by the domain experts and membership functions to derive a fuzzy output [9]. The final stage involves the defuzzification process which converts the fuzzy data back into crisp data for further processing [9].

Framing the fuzzy inference rules in such system involved a lot of trial and error strategies, and whenever there is an improvement, the rules need to be updated by the domain experts [9]. The fuzzy inference system used in predicting tsunami used five different editors to build, edit, and view the inference system [9]. The editors include fuzzy inference system (FIS) editor, membership function editor, rule editor, rule viewer, and surface viewer which were used to aid in managing the inference system [9]. Inputs to the system include earthquake (EQ), focal depth (FD), volcanic eruption index (VEI), landslide (LS), and height of the waves in the deep ocean [9]. The output of the system is either rare, advisory, or warning based on the validity of the occurrence of the tsunami [9].

Apart from this, the system using a sensor network to detect and mitigate tsunami has also been proposed based on the general recurrent neural network (GRNN) [11, 12]. This system uses two types of nodes, namely the sensor nodes to sense underwater pressure and the commander nodes that control the barriers with the help of a GRNN [12]. The proposed model in this paper is different from the above system since it uses the historical database to predict tsunami and uses a unique alert system to produce a tsunami alert. The proposed system can be combined with the other systems since it has a modular architecture to produce a real-time robust system as the tsunami is an unpredictable natural calamity and to improve the accuracy and performance of existing tsunami warning systems [13, 14].

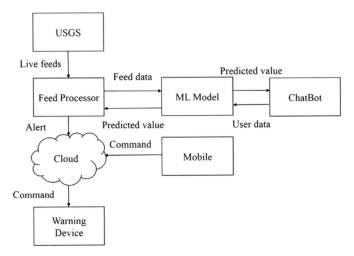

Fig. 1 Architecture of the proposed model

3 Proposed Model

The architecture of the proposed system is shown in Fig. 1. The proposed system is based on modern technologies such as machine learning, Internet of things (IoT), and cloud services. The system consumes the live feeds of earthquake data provided by the United Nations Geological Survey (USGS) which is updated every five minutes [15]. The data from USGS is then fed into the feed processor which monitors the live feed and formats the obtained feeds to the expexted format of the machine learning (ML) binary classification model. The ML model used in the proposed system is based on the random forests classification algorithm. The feed processor then communicates with the model trained on the NOAA dataset [16]. The ML model classifies the live feed data as Tsunamigenic or Non-Tsunamigenic based on whether the predicted feature named "tsunami" is set to 1 (tsunami = 1) or 0 (tsunami = 0). If the output is Tsunamigenic (tsunami = 1), the feed processor sends an alert command to the warning device, and the communication between them takes place through cloud IoT platform. The model also provides its service to the chatbot interface to improve the user experience by allowing the user to chat with the model for predicting the user-provided data.

3.1 Features of the Binary Classification Model

As mentioned above, the important characteristics of an earthquake that can cause a tsunami have been used as the fundamental criteria for choosing the features of the binary classification model. The features used in the model include magnitude, focal depth, and region of the epicenter, and if the epicenter is on the land, how long it has occurred from the nearest coastal point is also considered [17]. The dataset used for training the model has been taken from the NOAA Tsunami Warning Center's historical database of tsunamis, which holds tsunami records from 2100 BC [16]. The dataset is then filtered to contain only the records whose event validity was either 4 (definite tsunami) or 3 (probable tsunami) [5]. The negative data for the model was also prepared based on the above-mentioned criteria for Tsunamigenic earthquakes. The relationship between the different features used for training the model is depicted by Figs. 2, 3, 4, 5, and 6.

4 Random Forests Binary Classification Model

Random forests algorithm is an ensemble-based supervised classification algorithm [18]. As the name suggests, the algorithm works by building a set of randomized decision trees to avoid overfitting. This set of randomized decision trees can then be combined into a single strong learner which predicts the output based on the majority vote cast by the set of weak individual decision tree learners [18].

The general algorithm for random forests works as follows: If T is the training set to be fed into the model, then for every tree in the forest we choose a bootstrap sample from T depicted as $T(i)$ denoting the ith bootstrap [18]. The algorithm then selects a subset k from K, where k is the subset of total feature set K [18]. Each node in the tree that is being constructed is created based on the best split feature selected from k, and this process is repeated until the entire tree is constructed [18]. The main reason behind selecting a subset instead of the entire set of features is to avoid the

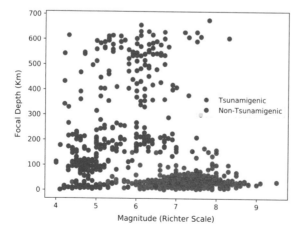

Fig. 2 Relationship between the magnitude and the focal depth of the data samples

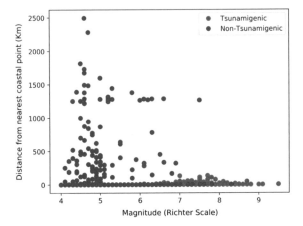

Fig. 3 Relationship between the magnitude and the distance from nearest coastal point of the data samples

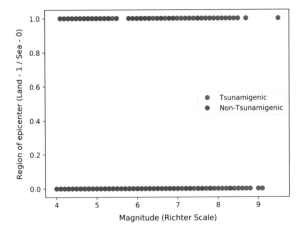

Fig. 4 Relationship between the magnitude and the region of epicenter of the data samples

overhead in finding the best split feature which in turn enhances the speed of the learning of the tree [18].

The bootstrapping technique followed in the above algorithm is known as bagging which helps in reducing the variance of the algorithm [18]. The reason behind choosing random forests algorithm for predicting tsunami is its higher accuracy rate in handling variations in data pattern and imbalanced datasets [19]. The algorithm gave better results compared to other classification algorithms with a proper balance between the bias and the variance of the model.

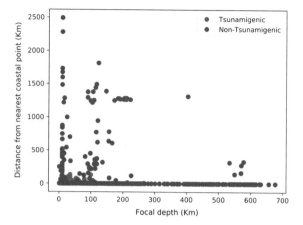

Fig. 5 Relationship between the focal depth and the distance from nearest coastal point of the data samples

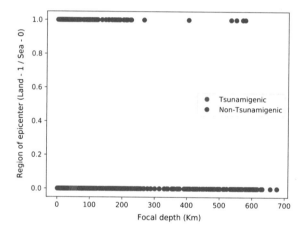

Fig. 6 Relationship between the focal depth and the region of epicenter of the data samples

5   IoT-Enabled Warning System

The warning system of Earthosys comprises IoT cloud platform, IoT-enabled warning device, and mobile device and consumes the service of the feed processor. The IoT-enabled warning device is deployed over the coastal regions. This device is locomotive and is controlled by a mobile through an application which specifies the direction in which the device must move. These direction commands are sent to the warning device through the IoT cloud platform as shown in Fig. 1.

When the feed processor detects a Tsunamigenic earthquake, an alert command is sent to the warning device through the cloud exactly in the same way as direction commands are sent. The device alerts the humans near the coastal region through a tsunami alert message. The IoT cloud platform consists of a variable named "tsunami" whose value is updated to 1 when an alert command is received. The warning device which monitors for a change in the value of this variable triggers the alert message to warn the surrounding environment immediately after the variable has been updated with the value of 1.

6   Evaluation of Machine Learning Model

In this section, the evaluation part of the machine learning model is discussed. The model was trained on all the valid and processed data samples. The overall accuracy achieved by the model on using random forests algorithm for predicting tsunami is 97.99% which represents the total number of predictions that were correct. Apart from this, the various evaluation metrics used in evaluating the model include confusion matrix, precision, recall, cross-validation score, and F1 score.

Confusion matrix provides a table view of true positives, true negatives, false positives, and false negatives between the actual result and the predicted result as shown in Table 1 [20]. After processing and formatting the NOAA provided dataset, there were a total of 995 valid tsunami samples for training the model. A total of 249 samples (25% of the entire dataset) of tsunami records which were not known to the model were used to test its performance, of which 244 samples were predicted precisely by the model as shown in Fig. 7. The samples that were predicted wrongly include 1 false positive and 4 false negatives. Precision provides an estimate of true positives out of samples that were predicted as true [21]. Recall or sensitivity provides an estimate of true positives out of total positive samples [21]. F1 score provides a harmonic average of precision and recall as shown in Table 2 [22]. As the accuracy score can sometimes be misleading due to bias and variance, the cross-validation score for the model is calculated. The cross-validation score estimated for the model is 97% which matched the accuracy score achieved by the model confirming the model's stability and the accuracy score's validity.

Under the experimental setup, the proposed system with the support of this model was able to produce an alert for a Tsunamigenic earthquake within 2.5 s through the IoT-enabled warning system. The model's output on some of the randomly selected

Table 1 Confusion matrix

Confusion matrix		Predicted class	
		Tsunamigenic	Non-Tsunamigenic
Actual class	Tsunamigenic	161	4
	Non-Tsunamigenic	1	83

Fig. 7 Bar chart depicting the actual samples and the correctly predicted samples in each class

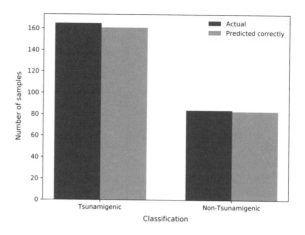

Table 3 Precision, recall, and F1 score

Class	Precision	Recall	F1-score	Samples
Tsunamigenic	0.95	0.99	0.97	165
Non-Tsunamigenic	0.99	0.98	0.98	84
Average/total	0.98	0.98	0.98	249

historic tsunami records is shown in Table 3, where 1 in the model output column denotes a Tsunamigenic earthquake and 0 denotes a Non-Tsunamigenic earthquake.

7 Conclusion

The Earthosys project combines the utilization of modern tools such as machine learning, IoT, cloud to provide a solution to a real-world problem. The use of machine learning in this project facilitates the learning process to be automated without any human intervention in framing the rules for prediction as the model learns on its own. The project architecture is framed in such a way that it provides a greater control over the tools that are being used in the prediction and warning system. The researchers framed the proposed model in such a way it utilizes the power of modern technologies to its best in protecting the humankind from the most detrimental disaster, tsunami. The proposed model is the initial version of this project, and many improvements are to be included in the upcoming versions.

Table 3 Evaluation result of the model on some of the randomly selected tsunami records

Tsunami records	Magnitude	Focal depth (km)	Region of epicenter	Distance (km)	Model output
Sumatra, 2004	9.1	30	Sea	0	1
South of Java Island, 2006	7.7	25.3	Sea	0	1
Kuril Islands, 2006	8.3	31	Sea	0	1
Solomon Islands, 2007	8.1	10	Sea	0	1
Samoa, 2009	8.1	15	Sea	0	1
Chile, 2010	8.8	35	Sea	0	1
Sumatra, 2010	7.7	20	Sea	0	1
New Zealand, 2011	6.3	5	Land	2.07	1
North Pacific Coast, 2011	9.0	29	Sea	0	1
New Zealand, 2016	7.8	15	Land	24.44	1

8 Future Work

The proposed model for prediction is trained based on the historical database of tsunami provided by NOAA [16]. In near future, the Earthosys binary classification model would be converted into a live-feed learning model where live-feed on earthquakes and its corresponding result would be fed into the model for training and to improve accuracy on all scenarios. Furthermore, the alert system which is currently a simple device that works with help of IoT and cloud can be fine-tuned into a more sophisticated robot.

References

1. National Oceanic and Atmospheric Administration. http://www.tsunami.noaa.gov
2. Indian Ocean Tsunami Information Center. http://www.iotic.ioc-unesco.org
3. Truong HV (2012) Wave-propagation velocity, tsunami speed, amplitudes, dynamic water—attenuation factors. In: Proceedings of world conference on earthquake engineering, pp 1–10 (2012)
4. International Tsunami Information Center. http://www.itic.ioc-unesco.org
5. Puspito NT, Gunawan I (2005) Tsunami sources in the Sumatra region, Indonesia and simulation of the 26 December 2004 Aceh tsunami. ISET J Earthquake Technol 42:4
6. Zaytsev A, Kostenko I, Kurkin A, Pelinovsky E, Yalciner AC (2016) The depth effect of earthquakes on tsunami heights in the Sea of Okhotsk. Turk J Earth Sci 25(4):289–299
7. All About Tsunamis. https://nhmu.utah.edu/sites/default/files/attachments/All%20About%20Tsunamis.pdf
8. US Geological Survey. https://www.earthquake.usgs.gov
9. Tayal T, Prema KV (2014) Design and implementation of a fuzzy based tsunami warning system. IJRET: Int J Res Eng Technol (2014)
10. Cherian CM, Jayaraj N (2010) Artificially intelligent tsunami early warning system. In: 12th international conference on computer modelling and simulation. IEEE Press, Cambridge, pp 39–44
11. Casey K, Lim A, Dozier G (2008) A sensor network architecture for tsunami detection and response. Int J Distrib Sens Netw 4(1):27–42
12. Casey K, Lim A, Dozier G (2006) Evolving general regression neural networks for tsunami detection and response. In: IEEE Congress on Evolutionary Computation. IEEE Press, Vancouver, pp 2451–2458
13. Angove MD, Rabenold CL, Weinstein SA, Eblé MC, Whitmore PM (2015) US tsunami warning system: capabilities, gaps, and future vision. In: OCEANS'15 MTS/IEEE. IEEE Press, Washington, pp 1–5
14. Le Mehaute B, Hwang LS, Van Dorn W (1971) Methods for improving tsunami warning system. In: IEEE 1971 conference on engineering in the ocean environment. IEEE Press, San Diego, pp 329–331
15. USGS Real-time Notifications, Feeds and Web Services. http://www.earthquake.usgs.gov/earthquakes/feed
16. National Geophysical Data Center/World Data Service (NGDC/WDS) (2018) Global historical tsunami database. National Geophysical Data Center, NOAA. https://doi.org/10.7289/v5pn93h7
17. Distance to the Nearest Coast. https://oceancolor.gsfc.nasa.gov/docs/distfromcoast
18. Random Forests. http://pages.cs.wisc.edu/~matthewb/pages/notes/pdf/ensembles/RandomForests.pdf
19. Rouet-Leduc B, Hulbert C, Lubbers N, Barros K, Humphreys CJ, Johnson PA (2017) Machine learning predicts laboratory earthquakes. Geophys Res Lett 44(18):9276–9282
20. Confusion Matrix. https://en.wikipedia.org/wiki/Confusion_matrix
21. Precision and Recall. https://en.wikipedia.org/wiki/Precision_and_recall
22. F1 score. https://en.wikipedia.org/wiki/F1_score

Computational Analysis of Differences in Indian and American Poetry

K. Praveenkumar and T. Maruthi Padmaja

Abstract Poetry is a verbal art which is motivating the human race from centuries. Indian authors' English poetry has its own signature in the world poetry. Though it has great significance when compared with Western poetry, very little work has been done in the areas of authorship affinities, classification, style similarity of poets, and comparative studies. In this work, we investigated style and semantic differences between Indian and Western poetry and also compared the poetry in terms of variation in the usage of words and stylistic features such as orthographic, syntactic, and phonetic features. To capture style and variation differences, we considered 84 style features that cover structural, syntactical, sound devices of poetry and computed TF-IDF values using a bag of word method; then, we computed cumulative TF-IDF value of each word across all poems and arranged the values in decreasing order of their cumulative TF-IDF value. Later, we applied ranks and used PCA, LSA, and Spearman correlation to find variance in usage of words by Indian authors' English poetry and Western poetry and style differences. We observed 40% semantic difference and 30% style difference between Indian authors' English poetry and Western poetry. Our comparative analysis says that the features that work well with one country poetry may not be necessary to perform well with other country poetry.

Keywords Style features of poetry · PCA · LSA · Spearman rank correlation

1 Introduction

In the wake of natural language processing, an extensive research has been reported on text in the form of prose, but very less work is reported on poetry. Poetry has its own specialty in its style of writing and careful use of vocabulary. Poetry analysis

K. Praveenkumar (✉) · T. Maruthi Padmaja
VFSTR Deemed to be University, Guntur, India
e-mail: praveenkumarkazipeta@gmail.com; praveen.kazipeta.kumar@gmail.com

T. Maruthi Padmaja
e-mail: padmaja.tu2002@gmail.com

© Springer Nature Singapore Pte Ltd. 2019
N. Chaki et al. (eds.), *Proceedings of International Conference on Computational Intelligence and Data Engineering*, Lecture Notes on Data Engineering and Communications Technologies 28, https://doi.org/10.1007/978-981-13-6459-4_13

can be done by using various aspects such as imagery, style, meter, and beauty [1]. Usage of these features varies from culture to culture, especially semantic difference which depends on culture. A good amount of work is going on the analysis of style, beauty, and imagery features of Western poetry [2–4].

However, Indian poetry in the English language has a great significance in portraying the culture and social issues of a time period, and it is well accepted by all over the world. Very less work is reported in the area of computational analysis of Indian English poetry. Indian poetry can be divided into pre-independent and post-independent poetry. In pre-independent poetry, the major focus is on themes such as patriotism, motivation, love, and god, whereas post-independent poetry focuses on contemporary and social issues [5].

Normally, poetry can be analyzed on its stylistic features, imagery, and semantic basis. Poetry is a composition of selective words which gives an in-depth sense of a particular theme; this poetry can be in lyrical, sonnets, haikus, etc., forms. Capturing features of a poetry latent structure and comparing Western poetry with Indian English poetry are a challenging task. Through our work, we try to explore this area and focused on the following points.

- Indian English poetry is how different with Western poetry in its style?
- How semantically and in usage of words differ?

2   Related Work

There is very little work which explicitly aims at measuring the difference between cross-cultural poetry. Johansson and Hofland [6] aimed to find the differences within LOB corpus; for this, they took 89 common words and calculated Spearman rank correlation. Biber [6] worked on different corpora to find the differences in sociolinguistic perspective.

2.1   Work on Analysis of Western Poetry

Kaplan and Blei [2] developed a poetry analyzer to find the style differences in American poetry by computing 84 features of American poetry and projected their similarities on two-dimensional spaces. They have used PCA to reduce the dimensionality. Kao and Jurafsky [1] compared the stylistic and content features of award winning and amateur poets with the help of style, affect and imagery features of poems and finds imagism is the most influential feature of poem to describe its beauty. Lamb et al. [7] compared poems for their novelty, meaning, and craft using consensual assessment technique where human judges will rate the poems by analyzing the features. For this, they used methods such as intraclass correlation, Kruskal–Wallis test, and open coding.

Lee [3] studied Chinese poems to find semantic parallelism by observing couplets pairs of adjacent lines in Chinese poems. He used semantic taxonomy which consists of 24 categories as features to find the semantic parallelism in Chinese poetry. Lee and Luo [8] explored the use of statistical methods to describe the parallelism in classical Chinese poems using graph-based clustering technique.

Al Falahi Ahemd et al. [4] worked on authorship attribution in Arabic poetry using Markov chain classifier. He has used character length, sentence length, and rhyme as features for his work. Kikuchi et al. [9] published their work on estimating artistic quality of haiku poems. They considered sound and meaning as features to find the artistic quality; for this, they used CNN model to construct quality estimation function.

2.2 Work on Indian Authors English Poetry

In this area, no computational work has been reported yet, but many linguistics people published their work on the comparison of Indian English poetry without any computational evidence. Rana [5] in her work described the rise of Indian English poetry before the independence and after the independence. How poets used their language, words to express various social, cultural and national issues. Pavan [10] published his work on pre-independence poetry of Indian English; in his work, he described the style of the poets and said that the poets are under the influence of Victorian and romantic strain stated that Indian English poetry is a mixture of Africa, America, British, and Australian poetry. However, for this work, there is no computational evidence.

2.3 Contribution of This Paper

Till now, there is no work reported, to the best of our knowledge that compares the Indian English poetry with Western poetry. Our work makes the following contributions: First, this paper introduces a new research problem on cross-culture poetry analysis and classification by doing a comparative study of poetry in terms of stylistic and semantic aspects. Second, there is no standard data set for computational analysis in Indian English poetry; we built a data set with around 600 Indian English poems that are penned by more than 30 authors. This data set can be used in the future to work on poetry classification and author style identification. Third, we have introduced a novel ranking method to calculate Spearman rank correlation of both Indian and Western poetries.

2.4 Outline of Paper

In Sect. 3, we discuss the data set we considered for both Indian and Western poetries and methodology used to find the style and semantic variance. Section 4 gives feature extraction method, and Sect. 5 gives a detailed analysis of results obtained. Section 6 gives conclusions and future directions to this work.

3 Data Set

Indian poetry: Indian English poetry is an oldest form of Indian English literature. It got name and fame across the globe. It projects various phases of national and cultural development from nineteenth century to mid-twentieth century. In Indian poetry, pioneer poets are Henry Derozio, Toru Dutt, S. C. Dutt; these poets paved a path and gave a new direction to the Indian poetry in English. Mostly their writings covered themes like Indian history, myths, and legends. Then, Swami Vivekananda, Sri Aurobindo, and Rabindranath Tagore portrayed the Indian culture, spiritualism, and Indian heritage. Next-generation poets like Nissim Ezekiel, K. N. Daruwalla, R. Parthasarathy, A. K. Ramanujan, Jayanth Mahapatra, and many others projected Indian landscape by adopting an own style of Indianness. For thorough analysis, we have considered 69 poems of Indian poetry.

Western poetry: For Western poetry, we considered Oxford Anthology of Modern American Poetry that covers the prominent American poets covering various time periods and styles of poetry that includes poets like Walt Whitman, Edna St. Vincent Millay, Ezra Pound, Anne sexton, and many others, and 59 selective poems are considered for experiment.

4 Feature Set Extraction

For feature extraction, we have used poetry analyzer program developed by Kaplan and Blei [2] to extract stylistic features and used R programming tool to generate term-document matrix using tm package. For experiment purpose, we considered these as two sets. And same set of experiments run on both sets.

4.1 Feature Set 1: Stylistic Features (84) of Indian and American Poetry

A poem can be characterized based on its style, imagery, affect, and semantic grounds. Poems style can be described using various features; we considered 84 features to describe style of poem, and they are categorized into three groups

Orthographic Features: These features describe physical structure of poem, which includes word count, number of lines per stanza, average line length, and average word length.

Syntactic Features: These features deal with the connection among parts of speech (POS) elements of poem.

Phonemic Features: In this category, rhyme and various forms of it are considered as the features.

4.2 Feature Set 2: Term-Document Matrix of Both Data Sets

We have computed term-document matrix (TDM) of poems, of both Indian and Western; while performing TDM, we have considered TF-IDF as weighting factor, and no stop words are used because we assumed that poetry is a verbal art and poets use the words carefully, so each word contributes to our goal and stemming performed to find the root word.

5 Experimental Results and Discussion

5.1 Experiment 1 on Feature Set 1: PCA Analysis on Stylistic Features

Principal Component Analysis: The very purpose of PCA is to replace a huge number of correlated variables with a less number of variables by retaining the much information of original variables. This small number of variables is called principal components; these components will preserve the most variance of the original variable set. The idea is to compare the PCA values of both American and Indian poetry to find the variance of the poetry and how many principal components are required to represent the maximum variance of poetry.

In Figs. 1 and 2, we have shown that the plot displays the scree test based on eigenvalues, and the red-colored dashed line suggests the number of components required. It shows approximately 9 principal components for Indian poems and 12 principal components for American poems required.

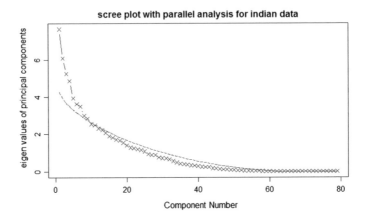

Fig. 1 PCA scree plot for Indian poetry

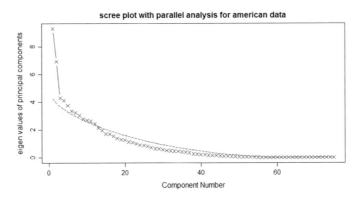

Fig. 2 PCA scree plot for American poetry

Table 1 PCA summary for Indian and Western poetry for three principal components

Item	PC1		PC2		PC3	
	Indian	Western	Indian	Western	Indian	Western
SS loadings	7.29	8.00	5.98	7.17	5.71	5.37
Proportion variance	0.09	0.11	0.08	0.10	0.07	0.07
Cumulative variance	0.09	0.11	0.17	0.20	0.24	0.24
Proportion explained	0.38	0.39	0.32	0.35	0.30	0.30
Cumulative proportion	0.38	0.39	0.70	0.74	1.00	1.00

Table 1 shows the principal component variance. In all aspects, Western poetry values are greater than Indian poetry.

Fig. 3 SVD diagonal values
scree plot for Indian poems

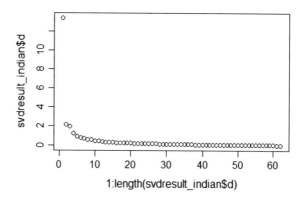

Fig. 4 SVD diagonal values
scree plot for American
poems

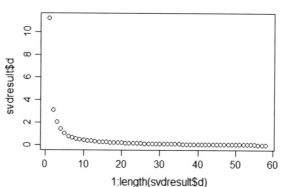

5.2 Experiment 2 on Feature Set 1: Latent Semantic Analysis

Latent semantic analysis will fetch the latent structure of the large set of data. It
is a dimensionality reduction method; it uses singular-value decomposing method.
It divides the given matrix into three matrices—two orthogonal matrices and one
diagonal eigenvalues matrix. The diagonal matrix will show the variance in the data;
by observing the scree plot of these values, we can say that how the data varies from
one data set to another. Here, in Figs. 3 and 4 of Indian poems' and American poems'
scree plots, we can observe very little variance in both curves. We call it as elbow
curve; at the curve point, it will say how many different concepts are present in the
data sets.

5.3 Experiment 3 on Feature Set 1: PAM Clustering

Partitioning around medoid clustering algorithm, here we applied this clustering
method on both data sets and observed the cluster information. This algorithm will

Table 2 Cluster information of Indian and Western poetry

Size of cluster		Max dissimilarity		Average dissimilarity		Diameter		Separation	
Indian	Western	Indian	Western	Indian	Western	Indian	Western	Indian	Western
14	13	1.75	2.23	0.94	1.25	2.42	2.77	0.65	0.72
17	18	1.25	3.29	0.79	1.23	1.69	4.52	0.49	0.72
15	22	1.28	2.18	0.80	1.00	1.81	2.49	0.49	1.22
11	4	2.1	1.11	1.15	0.62	2.64	1.33	0.86	1.27
5	1	2.7	0.00	1.34	0.00	3.86	0.00	1.5	6.73

find the random objects, and by iterating, it will update the medoid value. The result shows that Western poetry is clustered well than the Indian poetry; with this, we state that the Indian poems are with various distances than Western poetry (Table 2).

5.4 Experiment 4 on Feature Set 1: Spearman Rank Correlation

For all these 84 features, we got the values for poems in both data sets, then we calculated the cumulative value by computing row sum and arranged in ascending order, and then given rank for both poetry. Then, we computed Spearman rank correlation using Spearman function in R programming language. The correlation value is 0.73; it shows that 70% style features are similar to both Indian and Western poetry.

5.5 Experiment 5 on Feature Set 2: PCA Analysis on Word Occurrence

Figures 5 and 6 show more similarity, the result shows that for both poetry, to show the maximum variance 13 Principal Components are sufficient. But if we observe the Cumulative variance and SS loading it is more for Indian poems than Western poems this result is In contrast to the result of stylistic features PCA values. So it is evident that Indian poetry has some differences in semantically when compared with Western poetry (Table 3).

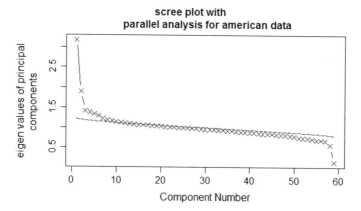

Fig. 5 Scree plot of PCA of American word occurrence data

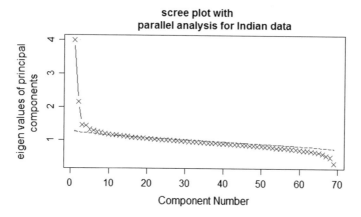

Fig. 6 Scree plot of PCA of Indian word occurrence data

Table 3 PCA summary for Indian and Western poetry for three principal components

Item	PC1		PC2		PC3	
	Indian	Western	Indian	Western	Indian	Western
SS loadings	3.17	2.52	2.24	2.02	2.18	1.94
Proportion variance	0.05	0.04	0.03	0.03	0.03	0.03
Cumulative variance	0.05	0.04	0.08	0.08	0.11	0.11
Proportion explained	0.42	0.39	0.29	0.31	0.29	0.30
Cumulative proportion	0.42	0.39	0.71	0.70	1.00	1.00

5.6   Experiment 6 on Feature Set 2: Latent Semantic Analysis on Word Occurrence

5.7   Experiment 7 on Feature Set 2: PAM Cluster Analysis

Figures 7 and 8 clearly show the variance in Western and Indian poems; the bends of the line shows that there is more variance in word occurrence in Indian poetry than the Western poetry. Silhouette Figs. 9 and 10 explains how good the data sets are clustered. The silhouette values shows that Indian poetry is better clustered than Western poetry.

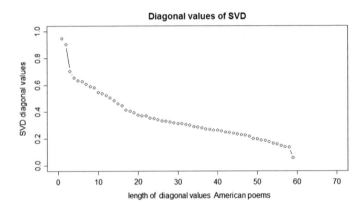

Fig. 7 SVD diagonal values scree plot for American poems

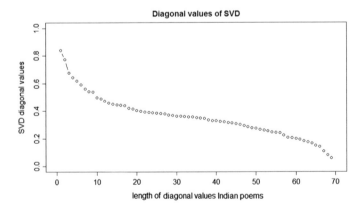

Fig. 8 SVD diagonal values scree plot for Indian poems

Fig. 9 Silhouette plot for American poetry

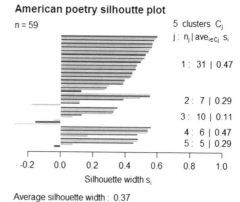

Fig. 10 Silhouette plot for Indian poetry

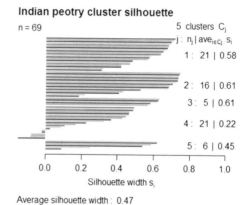

5.8 Experiment 8 on Feature Set 2: Spearman Rank Correlation

After computing the Term Document Matrix (TDM) of poems, of both Indian and Western, computed row sum to find the cumulative TF-IDF value for each word across all poems and arranged in ascending order so that we can get the most frequent words on top. Later, we have considered 1000 words from both Indian and Western poetry and computed the intersection of these 1000 words, we found 458 words as common, and for these common words, we have given ranking based on the TF-IDF value in respective classes of poetry. After we calculated spear man rank correlation for both poetry. The correlation coefficient value we got is 0.62; it shows that 60% of words are correlating in 50% of words of both poetries.

6   Conclusion and Future Work

Our method has shown the stylistic and semantic differences of Indian English poetry and Western poetry; the observations are though structural, the similarity between two poetries is high, but semantically, they are differing with 40% with Spearman rank correlation method. With this result, we can conclude that one method of computational analysis may not be suitable to describe all the data sets. We can further investigate that what are the other best feature extraction methods to describe the Indian poetry and classification methods to classify Indian and Western poetry.

Acknowledgements We profusely express our sincere thanks to Dr. C. Raghavendra Rao, Professor, University of Hyderabad, Central University, for his timely guidance in completing this work.

References

1. Kao J, Jurafsky D (2012) A computational analysis of style, affect, and imagery in contemporary poetry, pp 8–17
2. Kaplan DM, Blei DM (2007) Seventh IEEE international conference on data mining a computational approach to style in American Poetry, pp 553–558
3. Lee JSY (2008) Semantic parallelism in classical Chinese poems, pp 527–530
4. C. Using Markov, Authorship Attribution in Arabic Poetry, 2015
5. Rana S (2012) A study of Indian english poetry. 2(10):1–5
6. Kilgarriff A, Rose T, Kilgarriff A, Rose T (1998) Measures for corpus similarity and homogeneity
7. Lamb C, Brown DG, Clarke CLA (2016) Evaluating digital poetry: insights from the CAT. In: Proceedings of the seventh international conference on computational creativity, pp 65–72, June 2016
8. Lee J (2016) Word clustering for parallelism in classical Chinese poems P (W1) P (W2) Suppose N is the total number of tokens in the corpus, pp 49–52
9. Kikuchi S, Kato K, Saito J, Okura S (2016) Quality estimation for Japanese Haiku poems using neural network
10. Pavan GK (2014) English poetry and poets of pre-independent India. 2(2):57–62

Feature Selection for Driver Drowsiness Detection

Saurav Panda and Megha Kolhekar

Abstract Driver fatigue and sleepiness result in fatal accidents on road, many times. Drowsiness of driver is the major symptom of fatigue. Driver's attention can be checked from time to time to avoid such situations. This paper aims at detecting the driver's drowsiness using eye movement to detect open/close eyes. We give a generic design of the system that uses face detection, feature extraction and decision making through a trained model using support vector machine. The paper contribution is in comparing the performance of various feature extraction techniques and evaluating them using standard validation parameters. Features evaluated are the Canny edge, Local Binary Patterns, Histogram of Oriented Gradients (HOG) and Gabor filter bank, along with the normal gray image. They are evaluated for accuracy and "F1 score." Among All, HOG outperforms other methods and is a good choice for the application under consideration.

Keywords SVM · Feature Extraction · Drowsiness Detection · Classifier

1 Introduction

India, being a developing country, has one of the fastest rates of road network expansion and urbanization. This is accompanied by the ever-increasing demand for vehicles in our nation. Due to this, many accidents occur daily, in which hundreds of people lose the life. One of the main reasons for these accidents is driver fatigue or sleepiness [1]. In India, heavy vehicle drivers are usually sleep deprived due to the long journey, which results in driver fatigue and thus heavy vehicles account for

S. Panda (✉) · M. Kolhekar
Department of Electronics & Telecommunication Engineering, Fr. C. Rodrigues Institute of Technology, Navi Mumbai, India
e-mail: sauravgopinathpanda@gmail.com

M. Kolhekar
e-mail: megha.kolhekar@fcrit.ac.in

© Springer Nature Singapore Pte Ltd. 2019
N. Chaki et al. (eds.), *Proceedings of International Conference on Computational Intelligence and Data Engineering*, Lecture Notes on Data Engineering and Communications Technologies 28, https://doi.org/10.1007/978-981-13-6459-4_14

about 15% of the accidents. Impaired performance and subjective feel of drowsiness are the major symptoms of fatigue [2].

To avoid such accidents, one can use eye movements to check for the driver's attention. Eyes form one of the crucial features of the face, and eye images can be used to extract information about an individual. Eye movement, like open eyes and closed eyes, can be used to detect drowsiness of an individual, which can be accompanied by alerting system. This technology can help in avoiding a huge number of accidents and to save many lives.

One of the key challenges in handling this issue is to design a system, which is robust in nature and has accurate performance. The accuracy and robustness depend highly upon the reliability of the features. In this paper, we compare various features and classification methods for two major parameters, namely accuracy and response time. We calculate both training time taken for training the model and time taken for testing one frame.

2 Related Work

There is a lot of literature on recent developments in the design of robust systems for drowsiness detection. The main methods described are visual as well as non-visual methods. Eyepatch detector is one of the conventional methods.

Eye state detection algorithms using image processing can be briefly classified into heuristic-based algorithms and machine-learning-based algorithm. In the heuristic-based approach, a heuristic model is created using the dynamics of the problem. In [3, 4], an algorithm to detect open eye and closed eye using brightness and the numerical aperture of the iris is proposed.

In [5], drowsiness detection is classified into three major categories: image-processing-, ECG-, and artificial-neural-network-based techniques. Image-processing-based technique is further divided into template-based detection, blink measurement, and yawning-based detection. This paper concludes that ECG-based detection is more accurate but less feasible to implement, while AI-based technique needs more complex structure for drowsiness detection. Thus image-processing-based method is one of the most preferred approaches.

Multiple user identification along with eye tracking is proposed in [6]. With the help of three-dimensional imaging, that is an image with separate depth channel, it uses histogram of oriented gradients (HOG) based feature extraction and support vector machine (SVM). It claims a very high accuracy in tracking eyes. But its drawback is that the three-dimensional image sensor is unaffordable for practical purposes.

The machine learning-based approach involves training models to classify eye state using labeled data. In general, machine-learning algorithms are implemented using the following steps: feature selection, training the classifier, and testing the classifier. Due to multiple ways of designing this algorithm, choice of features and machine learning algorithm becomes crucial.

3 Features and Classifiers

This paper compares the following feature extraction techniques: Canny edge, local binary patterns, histogram of oriented gradients, and Gabor filter bank, along with the normal gray image.

3.1 Gray Image Feature

In colored images, each pixel consists of three values, each representing red, green, and blue (RGB) colors. These colored images are converted to a gray image. Each pixel in a gray image is the weighted sum of red, green, and blue pixel values.

$$Gray = 0.3R + 0.59G + 0.11B \tag{1}$$

A gray image feature vector is obtained from its two-dimensional representation by converting each pixel into a gray value and forming a one-dimensional array using (1). Gray is the gray value calculated for a given pixel with R, G, and B being red, green, and blue filter response of that pixel.

3.2 Canny Edge Image

The Canny edge features are obtained by performing Canny edge detection on the grayscale images to get a two-dimensional image, and then to get a one-dimensional feature array through flattening. Canny edge detection algorithm [7] has the following steps: Gaussian filtering, edge detection, non-maximum suppression, double threshold, and edge tracking by hysteresis. We have performed Canny edge detection using open-source module available in scikit-image [8].

3.3 Histogram of Oriented Gradients

HOG counts the histogram of gradients orientation in a localized portion of an image [9]. This feature descriptor is computed using following steps: gradient computation, orientation binning, descriptor block, block normalization. HOG features can be optimized by varying pixels per cell, cells per block, and number of orientations.

Initially, the gradient in x and y directions is computed using (2) and (3). g_x and g_y represent the gradient calculated in x and y directions, respectively.

$$magnitude = \sqrt{g_x^2 + g_y^2} \tag{2}$$

$$\theta = \arctan\left(\frac{g_y}{g_x}\right) \tag{3}$$

Image window is then divided into a small spatial region known as the cell. Size of this cell is given in pixels and varying the number of pixels per cell will vary it. Each cell has its own one-dimensional orientation histogram, which is calculated using the gradient magnitude. This divides the whole cell into a fixed number of orientation bins. To make HOG feature invariant to illumination, normalization is done on the pixels in given block.

3.4 Gabor Filter

Feature vector generated from the response of Gabor filter is widely used as feature descriptor. Multiple Gabor filters with different frequency and orientations are applied to the image, and overall response is considered as Gabor feature [10].

$$G(x, y, f, \theta) = e^{\left(\frac{-1}{2}\left[\frac{x'^2}{\delta_x^2} + \frac{y'^2}{\delta_y^2}\right]\right)} \cos(2\pi f x') \tag{4}$$
$$x' = x\cos(\theta) + y\sin(\theta) \tag{5}$$
$$y' = x\cos(\theta) - y\sin(\theta) \tag{6}$$

Gabor feature is calculated using (4). x and y are indices of row and column pixels of the image, respectively. f is the central frequency of the Gabor filter and indicates the orientation. In Gabor-based feature extraction, frequency and orientation dominate the response. Gabor filter bank can be used with varied combination to extract different information from the image. Deng et al. [11] discuss the use of Gabor feature to recognize a face.

3.5 Local Binary Patterns (LBP) Feature

The local binary pattern is simple texture-based feature extraction. It uses its neighboring pixel value to calculate the local patterns. Being computationally efficient and invariant to monotonic gray levels in the image, LPB is widely used as feature descriptor. Simple LBP descriptor considers 3×3 neighbor and threshold them. Decimal conversion of these neighboring values is considered as the value of centre element [12].

3.6 Support Vector Machine (SVM) Classifier

Support vector machine (SVM) is one of the most powerful classification algorithms. In this classification technique, n-dimensional space is formed where n is the number of feature use for classification. Value of each feature is used as co-ordinate for a respective plane. A hyperplane is selected in such a way that it has a higher margin of robustness. This ensures that there is less misclassification of data [13].

Figure 1 shows basic classification using support vector machine. SVM aims at finding the hyperplane which has largest minimum distance between the support vectors of given input classes. Samples which are closest to hyperplane are known as support vectors. Support vectors are used in calculating the optimal hyperplane.

4 System Design

In this section, we describe the work approach, methodology, and evaluation.

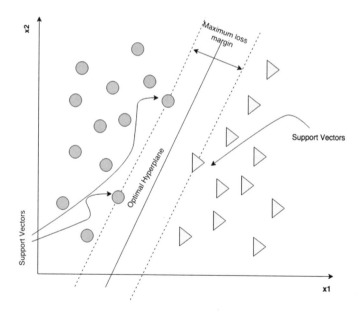

Fig. 1 Illustration of optimal hyperplane selection using Support Vector Machine (SVM)

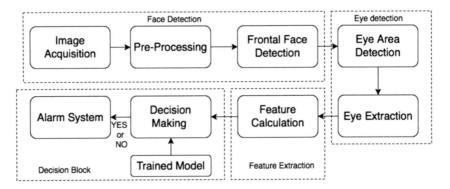

Fig. 2 Generic block diagram

4.1 Generic Block Diagram

Figure 2 shows the generic block diagram of a drowsiness detection system. It consists of four main subsections, which are face detection, eye detection, feature extraction, and decision-making.

The acquired image is first pre-processed to adjust the contrast and size of the image. This image is passed through a face detector to detect all the frontal faces present in the given image. Frontal face refers to face directly facing the camera. This frontal face is then passed to the eye detector block where both the eyes are detected. Both the eyes are cropped and passed to the feature extractor. A feature extractor extracts the feature vector based on the specified method. Feature vector generated is passed through the decision-making system, where it is compared with the previously trained model, to detect if the eye state should be classified as open or close.

4.2 Data Set

Analysis of various features is done on Closed Eyes in Wild (CEW) dataset provided by Tan [14]. This dataset contains 1192 subjects with closed eyes taken randomly from the Internet and 1231 subjects with open eyes collected form Labelled Face in Wild [14]. It provides with three different variations of this data. It has images with original resolution, resized images, and eye patches only. Given dataset extracts eye first by extracting the faces by using a Haar-like feature-based ensemble detector [15] and then localizes the eye using the discriminative pictorial structural model [16].

A dataset of extracted eye portion referred as eye patch is utilized to extract various features and analyze their performance. All these eye patches are resized to 24×24.

4.3 Pre-processing

Pre-processing stands for the process, which is done prior to performing the actual task to get the data in the prerequisite format. For drowsiness detection, the preprocessing step includes adjusting the gamma of the image, detecting face and eyes, cropping eyes, and resizing it.

For the images available in the CEW dataset, eye patch data set is used. It contains eye patch images, which are extracted and resized to 24×24. To modify the brightness level of this image, gamma adjustment with a gamma of 0.8 is applied.

4.4 Feature Extraction

The various feature like edges, histogram of oriented gradients and Haar-like features are extracted from the image. Grayscale images of 24×24 are flattened to form feature vector of length 576 for a given image.

To get Canny edge features, canny edge detection is applied to the image. This edge image is converted into a feature vector similar to the gray image. Figure 3 shows Canny edge images for open and closed eyes.

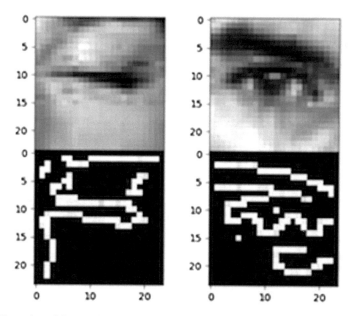

Fig. 3 Illustration of Canny edge feature image for closed eyes (left column) and open eyes (right column). Original images are shown on the top row with canny edge feature image at the bottom

A detailed analysis for HOG descriptor is done considering various parameters like cells per block (cpb), pixel per cell (ppc), and number of orientation (ori) bin. Four different variations for a number of pixels per cells are analyzed which are (2,2), (4,4), (6,6), and (8,8). Figure 4 shows the HOG feature descriptor with these variations, while orientation is set to 9 and block normalization of (3,3).

Various orientations were also considered with a constant pixel per cell as (8,8) and block normalization of (3,3). Feature descriptors with orientations 2, 4, 8, 9, and 16 are shown in Fig. 5. Figure 6 shows HOG descriptors with block normalization of (1,1), (2,2), (3,3).

By varying the various aspects of HOG descriptor, the size of the generated feature vector changes. This also affects the time taken for calculating the features. For a robust system, this time plays an important role in its selection along with the size of the feature vector. Gabor feature is calculated using 16 Gabor filter with varied orientation. Frequency and kernel size are kept constant. Similarly, for the LBP feature, 8 nearest neighbors are considered to calculate LBP image. The response

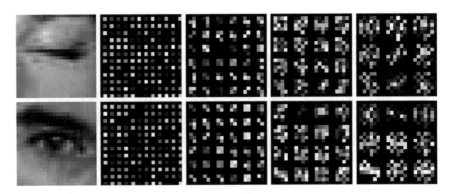

Fig. 4 Visualization of HOG descriptor with different pixels per cell value for closed eyes (top row) and the open eyes (bottom row). First column represents the original image. The pixel per cell values are (2,2), (4,4), (6,6), and (8,8) from left to right (excluding first column)

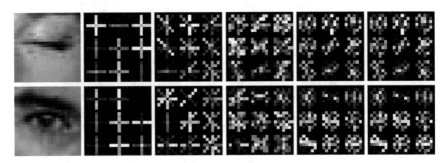

Fig. 5 Illustration of HOG descriptor with different orientations for closed eye (top row) and open eye (bottom row). First column represents the original image. The orientations are 2, 4, 8, 9, 16 from left to right (excluding first column)

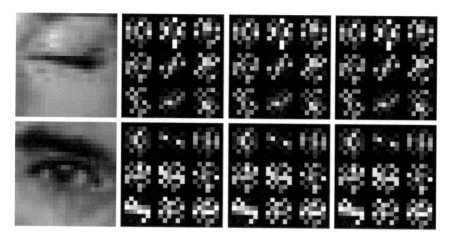

Fig. 6 Visualization of HOG descriptor with various block sizes (for normalization) for closed eye (top row) and open eye (bottom row). First column represents the original image. The cells per block values are (1,1), (2,2), (3,3) from left to right (excluding first column)

Table 1 Feature extraction time and feature vector size

Feature descriptor	Feature size	Extraction time (sec)
Gray	576	0.5400180
Canny edge	576	2.17103958
HOG (ppc=(2,2),cpb=(3,3),ori=9)	8100	58.1055879
HOG (ppc=(4,4),cpb=(3,3),ori=9)	1296	16.7180299
HOG (ppc=(6,6),cpb=(3,3),ori=9)	324	9.47713851
HOG (ppc=(8,8),cpb=(3,3),ori=9)	81	5.32388687
HOG (ppc=(8,8),cpb=(2,2),ori=9)	81	5.62500953
HOG (ppc=(8,8),cpb=(1,1),ori=9)	81	6.01100921
HOG (ppc=(8,8),cpb=(3,3),ori=2)	18	2.92491912
HOG (ppc=(8,8),cpb=(3,3),ori=4)	36	3.66687774
HOG (ppc=(8,8),cpb=(3,3),ori=8)	72	4.46093082
HOG (ppc=(8,8),cpb=(3,3),ori=16)	144	8.78107547
Local Binary Pattern (LBP)	576	0.945448
Gabor Filter Bank	576	0.7896180

of both Gabor filter bank and LBP are in 2 dimensions that are then converted into one-dimensional feature vectors.

Table 1 shows the time required to extract feature along with feature size for various feature descriptors.

4.5   Training the Classifier

Support vector machine (SVM) is used for classifying the image into open and closed eyes. A total of 3700 images are utilized to train the classifier, 1800 images are selected randomly from the closed eye images, and 1900 images from open eye images. The binary classifier is trained using SVM with label 0 for closed eyes and label 1 for open eyes.

The various parameters of SVM such as C value (loss margin) and kernel function are analyzed. SVM with C value ranging from 1 to 400 are trained, and their output response is analyzed. Four different kernel functions are analyzed which are linear kernel, polynomial kernel, radial basis function (RBF) kernel, and sigmoid kernel.

4.6   Model Evaluation

Trained model is evaluated on the basis of 5 statistical parameters, namely Accuracy, Recall, Precision, False Negative Rate (FNR), and F1 score.

For any binary classifier, the output instances can be represented by two-by-two confusion matrix known as contingency table (also called crossbar or cross tabulation). Figure 5 shows the contingency table for binary classifier. Samples that are positive and are classified by the classifier as positive are known as True Positive (TP). Samples that are negative and are classified as negative are called True Negative (TN). Similarly, falsely classified samples which are positive is known as False Positive (FP), and falsely classified negative samples are known as False Negative (FN).

$$\text{Recall} = \frac{\text{TP}}{\text{TP} + \text{FN}} \tag{7}$$

Recall is also known as the number of true positive rate which is estimated by (7).

$$\text{FNR} = \frac{\text{FP}}{\text{TN} + \text{FP}} \tag{8}$$

False negative rate (FNR), also known as false alarming rate, is given by (8). For a good classifier, FPR should be minimum.

$$\text{Accuracy} = \frac{\text{TP} + \text{TN}}{\text{TP} + \text{TN} + \text{FP} + \text{FN}} \tag{9}$$

Accuracy of a classifier is the rate at which it classified the sample to there respective classes correctly. Accuracy is given by (9). It is one of the widely used parameter to evaluate a classifier.

$$\text{Precision} = \frac{\text{TP}}{\text{TP} + \text{FP}} \tag{10}$$

$$F1_{score} = \frac{2 * Recall * Precision}{Recall + Precision} \qquad (11)$$

Precision is the measure to check how many samples are relevant given all correctly classified samples and is given by (10). Precision and Recall are together used to calculate F1 score of a classifier. The F1 score is the harmonic mean of Recall and Precision and is given by (11). Classifier having F1 score closer to 1 are known to perform better.

All these statistical parameters used to evaluate classifier in this system are explained in detail in [17–19].

5   Results

Histogram of oriented gradients (HOG) is analyzed in various configurations. Figure 7 shows the accuracy versus C parameter for SVM with different pixel per cell configurations. It is observed that the accuracy of the system increases as the number of pixels per cell decreases. At low C parameter value, (2,2) pixel per cell performs the best with the accuracy of 90%. (8,8) pixel per cell showed the least accuracy of 83% for low C parameter value. As the C parameter value increases, an accuracy of the system increases. (2,2) and (4,4) pixel per cell configuration showed the same accuracy for high C parameter value. From Table 1, it is observed that the size of the feature vector increases with decreasing pixels per cell. Thus (4,4) pixel per cell helps to get a perfect trade-off for this application.

Figure 8 shows the accuracy (in percentage) versus C-parameter variation for different orientation values. It is observed that the 4 orientations perform better than 2, 8, 9, and 16 orientations. For higher C parameter values, 8 orientations give the highest accuracy of 91.7%. All this orientations performance is analyzed by keeping

Fig. 7 Accuracy versus C-parameter value graph for 4 different configurations of pixel per cell (accuracy in percentage)

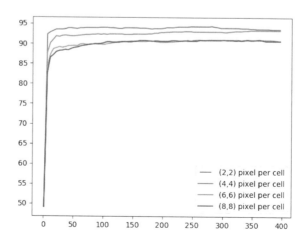

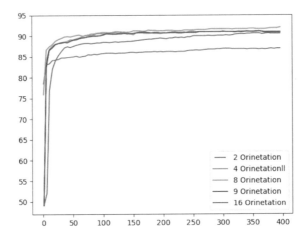

Fig. 8 Accuracy versus C-parameter value graph for 5 different orientations which are 2, 4, 8, 9, and 16 (accuracy in percentage)

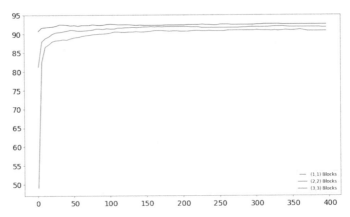

Fig. 9 Accuracy versus C-parameter value graph for 4 different configurations of pixel per cell (accuracy in percentage)

pixel per cell as (8,8) and block normalization of (3,3). For drowsiness detection, 8 orientations provide a perfect trade-off between accuracy and size of the feature vector.

Block normalization is performed to normalize the contrast in a given size of the block. Figure 9 shows the accuracy versus C-parameter curve for (1,1), (2,2), (3,3) cells per block. Table 1 shows a variation of time required to calculate features with different block sizes. It is observed that the accuracy and time increases decreasing cells per block.

Table 2 presents the statistical results obtained by training various features into the SVM model. HOG outperformed other feature descriptors in terms of accuracy and False Negative rate.

Table 2 Feature extraction time and feature vector size

Feature	Accuracy	FNR	Precision	Recall	F1-score
Gray	88.89	0.08	0.9174	0.8591	0.8873
Canny	84.178	0.17	0.84156	0.8487	0.8451
HOG	92.56	0.07	0.9277	0.9261	0.9269
LBP	84.61	0.15	0.8476	0.8505	0.8490
Gabor	88.28	0.11	0.8848	0.8848	0.8848

6 Conclusion

In this paper, we have compared various techniques for feature extraction of eye area in detecting an open eye or closed eye. The targeted application is driver drowsiness detection.

A trade-off between feature extraction time and accuracy is observed as expected. The time to extract the feature is directly related to the feature vector size of a particular method. The gray, Canny, HOG, LBP, and Gabor features are evaluated for accuracy and F1 score. Among all, HOG outperforms other methods and is a good choice for this application if the feature vector size is kept in moderation by changing appropriate cell sizes.

Future work involves hardware development for real-time drowsiness detection implementation. We intend to use Web camera along with raspberry pi which will continuously monitor the driver. Alarm signal will be activated if drowsiness is detected for a given time period.

References

1. Ministry of Road Transport and Highways (2018) Annual report on road accidents in India-2017 [Online]. Accessed 12 Apr 2018 [Online]. Available: http://www.indiaenvironmentportal.org.in/content/448147/road-safety-annual-report-2017/
2. Daza IG, Bergasa LM, Bronte S, Yebes JJ, Almazán J, Arroyo R (2014) Fusion of optimized indicators from advanced driver assistance systems (ADAS) for driver drowsiness detection. Sensors 14(1):1106–1131
3. Tabrizi PR, Zoroofi RA (2009) Drowsiness detection based on brightness and numeral features of eye image. In: Fifth international conference on intelligent information hiding and multimedia signal processing, 2009. IIH-MSP'09. IEEE 2009, pp 1310–1313
4. Tabrizi P, Zoroofi R (2008) Open/closed eye analysis for drowsiness detection. In: 2008 first workshops on image processing theory, tools and applications, Nov 2008, pp 1–7
5. Fuletra JD, Bhatt D (2013) A survey on driver's drowsiness detection techniques. Int J Recent Innov Trends Comput Commun 1(11):816–819
6. Kang S-J (2016) Multi-user identification-based eye-tracking algorithm using position estimation. Sensors 17(1):41
7. Canny J (1986) Computational approach to edge detection. IEEE Trans Pattern Anal Mach Intell 8:679–698

8. Van der Walt S, Schönberger JL, Nunez-Iglesias J, Boulogne F, Warner JD, Yager N, Gouillart E, Yu T (2014) Scikit-image: image processing in python. PeerJ 2:e453
9. Dalal N, Triggs B (2005) Histograms of oriented gradients for human detection. In: IEEE computer society conference on computer vision and pattern recognition, 2005. CVPR 2005, vol 1. IEEE, pp 886–893 (2005)
10. Kamarainen J-K (2012) Gabor features in image analysis. In: Image processing theory, tools and applications (IPTA). IEEE, pp 13–14
11. Deng H-B, Jin L-W, Zhen L-X, Huang J-C et al (2005) A new facial expression recognition method based on local gabor filter bank and pca plus lda. Int J Inf Technol 11(11):86–96
12. Ahonen T, Hadid A, Pietikainen M (2006) Face description with local binary patterns: application to face recognition. IEEE Trans Pattern Anal Mach Intell 28(12):2037–2041
13. Gunn SR et al (1998) Support vector machines for classification and regression. ISIS Tech Rep 14(1):5–16
14. Song F, Tan X, Liu X, Chen S (2014) Eyes closeness detection from still images with multi-scale histograms of principal oriented gradients. Pattern Recogn 47(9):2825–2838
15. Viola P, Jones MJ (2004) Robust real-time face detection. Int J Comput Vis 57(2):137–154
16. Tan X, Song F, Zhou Z-H, Chen S (2009) Enhanced pictorial structures for precise eye localization under incontrolled conditions. In: IEEE conference on computer vision and pattern recognition, 2009. CVPR 2009. IEEE, pp 1621–1628
17. Powers DMW (2011) Evaluation: from precision, recall and f-measure to roc, informedness, markedness and correlation. J Mach Learn Technol 2(1):37–63
18. Fawcett T (2006) An introduction to roc analysis. Pattern Recogn Lett 27(8):861–874
19. Majnik M, Bosnić Z (2013) Roc analysis of classifiers in machine learning: a survey. Intell Data Anal 17(3):531–558

Monitoring and Recommendations of Public Transport Using GPS Data

R. Gopinath and Gopal K. Shyam

Abstract Transportation has become part of our daily life. There are different modes of transportation like buses, taxis, trains and planes, but monitoring and recommending vehicles are still a challenge and that is why the public transports are less attractive and neglected by people. This influences the people to use their own vehicles which in turn increases the traffic. We propose a monitoring and recommendation technique based on GPS data set to overcome this problem. In metropolitan cities like Bengaluru, there are more than 5000 buses. It is difficult for a person to find the bus to his desired destination, because he is unaware of the bus status. This can be solved by intimating him about the status with the help of GPS. In other words, informing the passenger about arrival and departure of bus and also informing passengers on board about the current location can help in getting the updated status. In this paper, we propose recommendations based on GPS data to inform about arrival and departure and we use existing bus voice system to inform about current location. Handling such big data set is a problem; so, we use MapReduce techniques to solve this problem. This proposal reduces the waiting time of the passenger who uses public transport.

Keywords MapReduce · Transportation system · GPS data · Exact time of arrival (ETA) · Terrain compensation module (TCM)

1 Introduction

Public transportation is a part of human life; if a person in metro cities wants to travel from one place to another, he has to rely on public transport [1–5]. For example, in Bangalore there are more than 5000 buses covering the entire city every day, it is difficult for a person to find the bus to a particular place in time without the knowledge

R. Gopinath (✉) · G. K. Shyam
School of C and IT, REVA University, Bangalore, India
e-mail: gopinathr91@gmail.com

G. K. Shyam
e-mail: gopalkrishnashyam@reva.edu.in

© Springer Nature Singapore Pte Ltd. 2019
N. Chaki et al. (eds.), *Proceedings of International Conference on Computational Intelligence and Data Engineering*, Lecture Notes on Data Engineering and Communications Technologies 28, https://doi.org/10.1007/978-981-13-6459-4_15

of arrival and departure, so recommending the bus to the desired destination is the aim of this paper. If a person from a different place is commuting in a bus and he is unaware of the routes, in such cases the buses are to be equipped with bus voice and GPS. Based on GPS data, every location is announced by bus voice so that passengers are aware of every stop and can get down in his/her destination easily [6, 7]. And another aim of this paper is to lessen the waiting time of a passenger and recommend the exact location and time of arrival of any particular bus in a particular route so that in peak hours passenger gets correct information of the bus and can reach destination as early as possible. In current-day scenario, the bus voice and display systems are not based on GPS data and it is just a manual or time stamp-based. In this paper, we recommend an algorithm called exact time of arrival (ETA) which exactly tells us time of arrival of any particular vehicle (bus), and GPS data collected at data centres are to be managed, so we use MapReduce techniques.

The following are key contributions to this paper:

- This paper recommends advanced searching of buses based on GPS data set.
- It also gives an exact time of arrival (ETA) using algorithm.
- Using the GPS data, location can be identified and displayed.

2 Motivation

A. *Bus voice*

Nowadays, all buses are equipped with bus voice system, which announces current location and the next location. But the existing system is manually operated, wherein driver has to press button such that current location is announced from the database that is already stored sequentially; based on this information, the system announces the locations.

B. *Display system*

The recommendations cannot be made to each person as the number of commuters are more; hence, recommendations are made at public places (bus stop), as the information is displayed at stops based on GPS data and ETA. By this mechanism, the commuters are aware of current status and location of the vehicle [8–10]. The data stored in the GPS base are huge and need to be managed using big data tools such as HDFS and MapReduce; here, we use MapReduce model to manage data and NoSQL technique to retrieve required data from large data set; one such technique is the graph database. Here, we use graph database model to relate the entities, so that any query can be dynamically answered.

3   Overview of the Proposed System

As shown in Fig. 1, it consists of receiver, MapReduce framework, GPS database and a graph database. The receiver gets the information about the vehicle from tower and processes it in a MapReduce framework; then, it relates the entities with a graph database; then, the information is announced using a speaker; also, the information is processed and exact time is calculated and displayed.

Until this stage, it is not a big issue but to handle the data we need a MapReduce model which takes the key-value pair and outputs the value. Our MapReduceMeasure model is mainly based on MapReduce, which is designed as a generic design and programming model for processing and generating large data sets. MapReduce has two key operations: Map and Reduce. A data set user specifies a Map operation that takes key = value pairs as input to generate a set of intermediate key = value pairs, and a Reduce operation that takes all intermediate values associated with the same intermediate keys as inputs to generate a set of output values.

Even though sufficiently generic to perform many real-world tasks, the two-phase MapReduce model is best at generating a set of values based on the same key. The impact of one key on the values generated by another key is difficult to evaluate in the current model.

Map: (key1; value1)! Set [key2; value2];

Reduce: (key2; Set [value2])! Set [value2];

The map output will be the input for reduce phase, and finally the large data set will be partitioned and merged accordingly.

A.   *Graph database model*

The data being stored have to be managed without using a SQL query technique as it is a time-consuming method; hence, we use a graph database model which relates the entities in a data, so that it would be an efficient way to identify and retrieve the

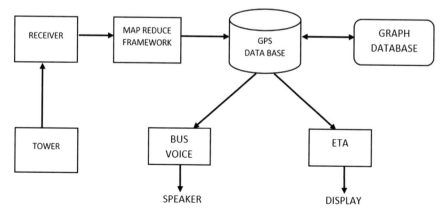

Fig. 1 Overview of recommendation system

required information. For instance in a bus, database and the information regarding bus such as number and status are related using directed arrows.

B. *Hot spot scanning*

This technique helps in identifying the vehicle in a Map; each time when a vehicle traverses the path it is marked as hot spot and number of times that vehicle traverses in same region for more than one time; then, it is a hot spot zone for that vehicle. By this, the traffic flow can be managed. There is an algorithm for marking hot spots.

Algorithm 1: Hot spot Marking

Input: Frequent surface TCM(x,y,z), F=(count(x,y)).
Output: Number of hot spot zones, range of each hot spot zone, representative point.
Step 1: If (n<range || F==0) return;
Step 2 : Perform simple division (TCM)

> *F=F/4;*
> *For i->0 to 3*
> *Ci= sumcount (sub_TCMi)*
> *remove TCM;*
> *N=N-1*

Step 3: For i->0 to 3

> *If Ci !<F*
> *Add sub_TCMi*
> *Hot-spot-scan (sub_TCMi,F,Lock)*
> *N = N+1*

This algorithm is used to mark number of hot spots and range of each hot spot, wherein this algorithm F may be the frequent surface where vehicle travelled and TCM is the trajectory of mobile trace vehicle in space.

C. *Exact time of arrival*

The intension here is to notify the passenger with exact arrival and departure of vehicle; hence, the estimated time can be given using the amount of time taken by a vehicle to reach a particular place considering its average km/ph. But when in extreme condition such as traffic and other cases, it is difficult to predict arrival and departure so we propose ETA algorithm which exactly tells the arrival and departure time.

Here, we consider the movement of vehicle in every fraction of second. For instance, if a vehicle V travels from A to B (100 m), it takes 10 s with a speed of 40 kmph; then, if it takes 15 s for next 100 m, so difference is calculated and an exact time can be predicted.

Similarly, the reset of the distance is calculated using same mechanism and exact time can be displayed at the stops.

Algorithm 2 is used to find exact arrival time of bus. Let v be the vehicle and A, B the source and temporary destination, respectively. Each time when the vehicle travels from A to B, the time taken by it is calculated; when v reaches B, the time is noted and B is made the new source; and it goes on till the vehicle reaches the desired destination.

Algorithm is as follows: let v be the vehicle and A, B be the source and temporary destination.

Algorithm 2: Exact Time of Arrival

*Step 1: For any given time **t** the distance travelled;*

*From A-> B is x: **A->B=x;***

*Step 2: The average speed of vehicle: **v=p;***

Now when vehicle reaches B;

*Step 3: **Value=x;***
Step 4: B=temp;
Step 5: B->new source;
Step 6: B->next temp destination;
Step 7: Value=y;
Step 8: Difference between A->B and B->t==z
Step 9: Return z;

4 Conclusion

The above-proposed work helps in reducing the waiting time of any passenger at the stops, without having the knowledge of arrival and departures. The proposed ETA helps in calculating the exact time of arrival of the vehicle and its current location, so that the passenger is aware of the current status of the vehicle. Any person on board can know the current location, as the vehicles are GPS-enabled; as in older system, he may not rely on driver to press the button and announce the current location; instead, the proposed system takes the GPS data and announces the location for passengers' convenience.

The hot spot scanning helps in monitoring the traffic flow and identifies the vehicle in the map, so that other vehicles in the same route can have a clear knowledge about the traffic and passengers as well. The MapReduce and graph database concepts help in managing the large data set and querying the required information. The simple concept with a touch of technology helps people in finding the desired vehicle to reach the correct destination.

References

1. Xiujuan Xu, Jianyu Zhou Yu, Liu Zhenzhen Xu, Zhao Xiaowei (2015) Taxi-RS: taxi-hunting recommendation system based on taxi GPS data. IEEE Trans Intell Transp Syst 16:1716–1727
2. Yuan J, Zheng Y, Xie X, Sun G Driving with knowledge from the physical world. In: 17th ACM SIGKDD international conference on knowledge discovery and data mining, pp 316–324
3. Zhang D, He T, Liu Y (2014) A carpooling recommendation system for taxicab services. IEEE Trans. Emerg Top Comput 2(3):254–266
4. Thomas G, Alexander G, Sasi PM (2017) Design of high performance cluster based map for vehicle tracking of public transport vehicles in smart city 2017. In: 2017 IEEE Region 10 symposium (TENSYMP), pp 1–5
5. Dow C-R, Chen H-C, Hwang S-F (2015) A hotspot aware taxi zone queuing system. In: 2015 International conference and workshop on computing and communication (IEMCON) pp 1–4
6. Rathod R, Khot ST (2016) Smart assistance for public transport system. In: 2016 International conference on inventive computation technologies (ICICT) vol 3, pp 1–5
7. Subadra KG, Begum JM, Dhivya H (2017) Analysis of an automated bus tracking system for metropolitan using IoT. In: 2017 International conference on innovations in information, embedded and communication systems (ICIIECS) pp 1–5
8. Gao R, Zhao M, Ye T, Ye F, Wang Y, Luo G (2017) Smartphone-based real time vehicle tracking in indoor parking structures. IEEE Trans Mob Comput 16(7):2023–2036
9. Saini A, Chandok S, Deshwal P (2017) Advancement of traffic management system using RFID. In: 2017 International conference on intelligent computing and control systems (ICICCS), IEEE conferences, pp 1254–1260
10. Papola A, Tinessa, F, Marzano V, Mautone A (2017) Quantitative overview of efficiency and effectiveness of public transport in Italy: The importance of using ITS. In: 2017 5th IEEE international conference on models and technologies for intelligent transportation systems (MT-ITS), pp 895–900

A Study of Sentiment Analysis: Concepts, Techniques, and Challenges

Ameen Abdullah Qaid Aqlan, B. Manjula and R. Lakshman Naik

Abstract Sentiment analysis (SA) is a process of extensive exploration of data stored on the Web to identify and categorize the views expressed in a part of the text. The intended outcome of this process is to assess the author attitude toward a particular topic, movie, product, etc. The result is positive, negative, or neutral. These study illustrated different techniques in SA approach for extracting and analytics sentiments associated with the polarity of positive, negative, or neutral on the topic selected. Social networks SA can be a useful source of information and data. SA acquires important in many areas of business, politics, and thought. So, this study contains a comprehensive overview of the most important studies in this field from the past to the recent studies till 2017. The main aim of this study is to provide full concept about SA techniques and its classification and methods used it. Also, we give a brief overview of big data techniques and its relation and use in SA field. Because the recent period has witnessed a remarkable development in the use of Big Data (Hadoop) in the process collection of data and reviews from social networks for analysis.

Keywords Big data · Classification · Challenges · Sentiment analysis · Social media · Twitter

1 Introduction

In present days, most of the people are expressing their feelings, opinions, and sharing their experiences, using the Internet and the social networks. This usually leads to communicate massive amount of data using the Internet. But most of these data are useful when analyzed; for example, most industrial companies and election cam-

A. A. Q. Aqlan (✉) · B. Manjula
Department of Computer Science, Kakatiya University, Warangal 506009, Telangana, India
e-mail: ameenaqlan218@gmail.com

R. Lakshman Naik
Department of Information Technology, Kakatiya University, Warangal 506009, Telangana, India

© Springer Nature Singapore Pte Ltd. 2019
N. Chaki et al. (eds.), *Proceedings of International Conference on Computational Intelligence and Data Engineering*, Lecture Notes on Data Engineering and Communications Technologies 28, https://doi.org/10.1007/978-981-13-6459-4_16

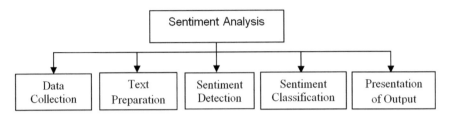

Fig. 1 Sentiment analysis process steps

paigns rely on knowing the opinions of people through communication sites and see whether they are positive, negative, or neutral. The SA has emerged because of the huge information exchange on the Internet. The SA idea was first proposed by Nasukawa [1]. Firstly, the SA is used for natural language process (NLP) [2], which analyzes opinions, feelings, reactions of people and writers on the Internet through social networking sites and business sites about the many products and services. Sentiment analysis is a broad field for many researchers and can also be called opinion mining; because it helps to classify ideas and opinions as positive, negative, or neutral. SA is a textual study, which is widely used on reviews and surveys in the Internet and social media. It handles responses and customer feedback on commercial sites to know their acceptance or rejection of a product; this helps to improve the sales of the company as it tells the choice of a customer. With the explosion of different opinions through social networking sites, new ideas were generated by systems, politicians, psychologists, manufacturers, and researchers to analyze them to implement the best decisions. Sentiment analysis has a high efficiency using NLP, as statistics, and machine learning approaches to extract and define sentiment content in a text unit.

2 Sentiment Analysis

Sentiment analysis is becoming very important to study growing opinions faster and faster within social media and other sites, The huge explosion in information in recent years in the sites of communication, air traffic and alternative markets, all this huge amount of information cannot be controlled and analyzed used the traditional way, so the scientists and researchers developed a high-efficiency techniques to deal with this data. This requires the SA to process data and know its polarity to determine the right decision. SA involves five steps to process data; those are data collection, text preparation, sentiment detection, sentiment classification, and presentation of output [3] as shown in bellow (Fig. 1).

2.1 Data Collection

The data collection is the first step in sentiment analysis. The collection of data from sources like user groups, Twitter, Facebook, blogs and commercial website such as amazon.com and alibaba.com, etc. This data cannot be analyzed using traditional methods like scanning, text analysis, and language processing, which is used for extraction and classification. Wei Xu [4] and Tapan Kumar [5] proposed a certain method for a task of paraphrasing and gathering the tweet data called Twitter Streaming API.

2.2 Text Preparation

Text preparation examines the data before analyzing it. Some reviews and conversations in the communication sites contain offensive and inappropriate words, so they are examined and preparation to be the result more reliable analysis. This process selects the contents that are not related to the analysis and then removes it. Objective of the process is the removal of spam and inappropriate reviews before sent to automated analysis. In this case, we can use part of speech (POS) technique which are used for text preparation before analysis [6, 7].

2.3 Sentiment Detection

Sentiment detection is the process of finding the sentiment newline expressed in a review by using machine learning technique or NLP technique; these are also called opinion mining (OM) new line and sentiment analysis. Sentiment detection consists of the examination of phrases and sentences extracted from reviews and ideas. All the sentences containing self-expressions like beliefs, opinions, and abuse are retained. Many research studies in this field included different methods of detection, like Lakshmish Kaushik [8]; one of the recent studies propose a system for automatic sentiment detection in natural audio streams by using POS technique.

2.4 Sentiment Classification

Sentiment Classification is a task to extraction and classification the text whose objective to classify according to a polarity of the opinion it contains (pang 2002), e.g. positive or negative, good or bad, like or dislike. Sentiment classification contains multiple techniques, and it is classified into three main techniques, namely machine learning approach, hybrid techniques approach, and lexicon-based approach [9] and

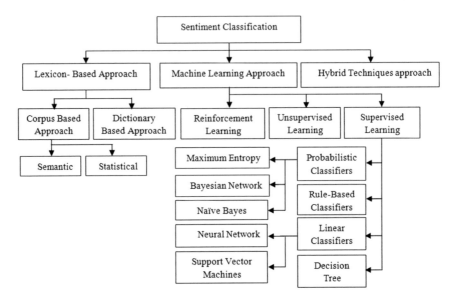

Fig. 2 Sentiment classification techniques

[10]. Presently, Naive Bayes technique and support vector machines (SVMs) are more popular and used for sentiment classification. These techniques can improve an accuracy of classification of Tweets, such as Ankur Goel [11] use Naive Bayes for Sentiment Analysis Tweets in high speed. Therefore, sentiment analysis has a great deal of research in this field and found that many applications and improvements have occurred in sentiment analysis in recent years. In this study, we will clarification most of the research in this area. These articles cover most of the divisions and classifications widely in SA fields. Sentiment classification techniques are discussed with more emphasis on most details and related points and originating references. In Fig. 2, we will illustrate all techniques which are currently used in sentiment classification until 2017.

2.4.1 Lexicon-Based Approach

Multiple words are used to classify sentiment and use positive words for the desired things, while using negative words for undesired things. So, lexicon-based approach relies mainly on finding opinion lexicon, which is used for text analysis. There are two methods according to lexicon-based approach. The first one is corpus-based approach, and the second one is dictionary-based approach.

Corpus-Based Approach: The corpus-based approach starts with a seed list of opinion words and then finds other ideas from the words in a large corpus to get opinions from certain directions. In another meaning, most methods rely on grammatical patterns or that occur together with the seed list of opinion words to find other words from a large corpus. Hatzivasiloglu [12] is one of the most important methods to represent the corpus-based approach. The first step to start was to create a seed list and use it with a wide range of language restrictions to be able to identify additional words including their orientations. To implement corpus-based approach, we use two different approaches: statistical approach and semantic approach as illustrated in the following.

i. **Statistical approach**: It is used in many applications that have a relation in the field of SA. The famous of them is the one that can detect the manipulation of the review by conducting a statistical test of randomization which is called runs test [13].

ii. **Semantic approach:** It gives values to sentiments while relies on more than principle to calculate the affinity and similarity of different words. The basis of this principle is to support the Sentiment value in the words and words close. WorldNet [14].

Dictionary-Based Approach: Dictionary-based approach presented a comprehensive strategy for the dictionary-based approach. In this famous strategy, a small group of words is hand-picked with known trends [15, 16]. Then, we come to plant this group of words by searching in the known approach corpora thesaurus [17] or WorldNet [18] for all synonyms and antonyms. The new words that are found are added to the seed list, and then, the next repetition begins. This repetitive process continues and stops only when there are no new words.

2.4.2 Machine Learning Approach

Machine learning approach is used to solve the problems related to text classification that contain syntactic or linguistic features. Whilst lexicon-based approach is used to extracting sentiment from text, it depends on a sentiment lexicon; the collection of known and pre-compiled sentiment terms in Machine Learning algorithms divided into Reinforcement Learning [17], Unsupervised Learning, and Supervised Learning.

Reinforcement Learning Technique: Its entirety indicates how to make an optimal decision an important technique that differs relatively from its counterpart unsupervised learning. This technique is highly concerned with improving the efficiency of text classification to show that the reinforcement learning technique is important and prominent.

Unsupervised Approach: It is a unique type of machine learning algorithm and is used in most cases to draw and diverse inferences of data; these groups of data consist of input data without any labeled responses. It is used when it is impossible to obtain labeled training documents.

Supervised Learning: It is a type of machine learning approach that uses a data set called training data set to make predictions. These data set contain input data as well as response values. In supervised learning methods, it makes use of a large number of assorted training documents.

Probabilistic Classifiers: It uses many models for classification. There are several types of mixture models; each mixture model must be an integrated mixture component. Each type of this mixture acts as generative and can support the particular taking term for this component or other; this approach is called generative classifier.

   i. *Maximum Entropy Classifier*: The maximum entropy classifier is a kind of classifications normally used in NLP, speech, data, and addressing problems. Maximum entropy is also probability distribution estimation; it is an important and famous technique widely used for a variety of natural language tasks, such as language modeling, part-of-speech tagging, and text segmentation. The underlying principle of maximum entropy is without external knowledge.
  ii. *Bayesian Network classifier:* The most important assumption of the Bayesian network classifier is a set of variables, each variable containing a limited set of mutual cases. It is independent of the features. The real assumption is a suggestion intended to have all the features which are completely dependent. This certainly leads to a certain model of Bayesian network which is a guided graph and represents a random contract.
 iii. *Naïve Bayes:* Naïve Bayes is the most popular method for text classification recently. Naïve Bayes classifier model computes the back probability of the class, based on the division of words in the adopted document.

Rule-based Classification: Rule-based classification is used for any scheme that constructs the classification according to rules IF and THEN. Linear classifier is a vector of real-valued numerical input features.

Linear classifier: is a decision based on the value of a linear combination of characteristics. An object's characteristics are also known as feature values and are typically presented to the machine in a vector called a feature vector; linear classifier is divided into two methods; those are:

   i. *Neural Network:* Neural network is a continuum of algorithms based on the recognition of the relationships inherent in several sets of data using a process similar to the way the human mind works.
  ii. *Support vector machine:* SVM is used to analyze datasets for classification and regression analysis, it is a machine learning algorithm that leads to process data automatically.

Decision Tree Classifiers (DTC): These are used for classification. It's aims to divide large data into small groups to be easy control, the DTC use multi-values of attributes, and features of the data to appear a class label discrete prediction. It is a fairly simple technique and widely used in sentiment analysis field.

2.4.3   Hybrid Techniques Approach

Hybrid techniques approach is a combination of multiple computational techniques which provide greater advantages than individual techniques and improve sentiment (data) analysis. Use of this technique is very convenient for many because it combines two or more technologies, so it shows much better results than other methods.

2.5 Presentation of Output

The main objective of analyzing a huge amount of data is to convert unstructured text into useful information and then to display it through charts such as a graph, line graph, and bar graph.

3 Background

SA is contextual mining of texts; it identifies the sentences and subjective information to classify opinions according to polarity. Sentiment analysis studies people's feelings, opinions, assessments, and attitudes toward many services, issues, events, and organizations [17]. SA is not only applied to the commercial product reviews; it can also be applied to all types of social communication sites and stock markets. Three topics work under the umbrella of sentiment analysis emotions detection, building resources, and transfer learning.

Emotions detection is a recent field of research that is closely related to SA. The aim of SA is to detect positive, negative, or neutral, feelings from the text, whereas emotion detection aims to detect and recognize types of feelings through the expression of texts, such as anger, disgust, fear, happiness, sadness, and surprise. Building resources is a lexicon; it is a vocabulary that is used to express an opinion according to the polarity either positive, negative, or neutral. Transfer learning is considered as the transfer of knowledge from one learned task to a new task in machine learning. The text classification according to the following criteria is as follows: The first standard is the polarity of sentiment into (positive, negative, or neutral). The second standard is the polarity of the outcome that applies to most political articles and medical facilities for managing disease data as follows [19]:

- Use agree or disagree, e.g., political debates [20];
- Criteria good or bad [21];
- Pros and cons: The meaning of this is either positive or negative [22];

in the following figure, will clarify steps of Sentiment Classification and related works in SA.

Fig. 3 Knowledge discovery and pattern recognition architecture

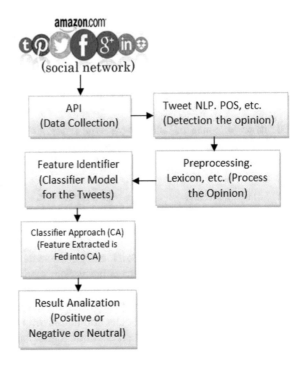

3.1 Knowledge Discovery and Pattern Recognition Architecture

Social networking sites are many and full of useful data; however, there are important and credible data and useless data (not useful). The reliability data is usually found in cultural sites or shopping sites because the customer may have experience in dealing with commercial sites. So, we have illustrated the kinds of techniques and classifications within the field of sentiment analysis and how to extract and manipulate data to reach reliable results. This section will illustrate a knowledge discovery and pattern recognition architecture, Fig. 3.

3.1.1 Social Network

The Internet is the right environment and the main source for most information and ideas that are shared by users. It is a resourceful place with respect to sentiment information. By following a lot of ideas and articles, we found that people prefer to publish their content through various online social media, such as forums, microblogs, or online social networking sites. Choosing data source is the first step in SA. All communication site a fertile environment for data collection; most of the data is useful, and often, there is abusive data which not useful, these data are excluded automatically during analysis.

3.1.2 Data Collection

Application program interface (API)—this is the proposed system to extract the data and download. It is characterized by research for hashtags, main keywords, and other classifiers simultaneously [23]. API is widely used to collect reviews by researchers and interested companies, but now we can use different Application program to collect data such as Hadoop flume, Shirahatti [21].

3.1.3 Natural Language Processing

Natural Language Processing: It deals with all human languages whether written or oral to process and apply. This is the main propose of NLP. NLP refers to part of text; part of text includes verbs, adjectives, and nouns.

We may get unorganized data in this case and need further processing using one of the algorithms (part of speech or N gram). Part of Speech can divide the sentences to small words, and each word has a meaning; for example, in English language each small word can have a distinctive meaning and this comes according to its use and functions, and is categorized into several types or small parts of speech such as noun, pronoun, verb, adverb, adjective, conjunction, preposition, and interjection.

N Gram, An n gram is simply a sequence of tokens. In the context of computational linguistics, these tokens are usually words, though they can be characters or subsets of characters. The n simply refers to the number of tokens. n gram is used for word sequence itself or predictive model that assigns it a probability. The gram is a combination of letters; n gram refers to divide the sentence into several parts (count the word); for example, "Friday will be holiday"—this is contains 4 gram; "tomorrow is holiday"—this is contains 3 gram. Named-entity recognition (NER), using to divides comments or the tweet into smaller parts this parts each one containing two words.

3.1.4 Preprocessing

Preprocessing relies mainly on finding opinion lexicon, which is used for text analysis; it classifies the words either positive, negative, or neutral.

3.1.5 Feature Identifier

Feature identifier: It is to identify and classify the entities being referred to as tweets. Final stage SVM classifier is trained in order to obtain the tweet's label.

3.1.6 Classifier Approach

Classifier Approach: It is used to analyze datasets for classification and regression analysis. It classifies the opinion positive, negative, or neutral in a final stage.

3.1.7 Result Analyzation

Result Analyzation or sentiment analysis commonly uses several ratings to express the abundance of feelings or versa. Sentiment is evaluated through the use of stars; some Websites require the evaluation of their material by using stars, and we will illustrate the rating of stars commonly used as follows [24] and [25].

- Positive +2 or 5 stars
- Rational positive +1 or 4 stars
- Neutral 0 or 3 stars
- Rational negative –1 or 2 stars
- Negative –2 or 1 star

4   Related Work

The purpose of this study is to give a clear conception of most techniques in the field of sentiment analysis, where it is easier for new researchers to benefit from it. As mentioned many techniques of analysis, we will clarify some studies and research in recent years that dealt with this area; this paper also covers a wide field of sentiment classification technique and approach in SA field. Lexicon-based technique aims to extract and collect data from social network such as Twitter [25], Facebook, etc. by use API Graph to collect and load all the target data for analysis, and examine all words that do not represent an emotional value or feature, then created a list of words and analyzed them, that would be used in all cases, these shown positive results in predict the sentiment behind a status post on Facebook by use lexicon-based approach with high efficiency.

Machine learning approach is not limited to the analysis of data in social media, Where used to know the driver's sense at the moment of leadership. One of them sought to generate and know the rules of the cognitive deviations of the drivers directly from the place of the driving simulation environment. Through this study, the eye movements of the drivers were taken using a simulated device [26].

Dictionary-based approach is used with high efficiency in the field of SA. Seongik Park build thesaurus lexicon characterized in clearly and credibility [27], Where build this approach through three online dictionaries to gathering thesauruses based on the seed words, and sought to stores the real words which can be trusted into the thesaurus lexicon in order to improve the reputation and credibility of the thesaurus lexicon, and prove it a prominent lexicon.

Use the DBA to build thesaurus lexicon. The purpose of this is to increase the availability of tweet and review for the sentiment classification without the need to use human resource. However, accuracy obtained was slightly increased.

Ishtiaq Ahsan has build a methodology for reviewing opinions and detecting spam through a well-known learning method (active learning and supervision) with the use of all data including real and fabricated show us very promising results while conducting several different experiments [21]. The results have shown that detection method is very effective and promising. The use of this technique is very convenient for many because it combines two or more technologies, so it shows much better results than other methods.

Seongik Park [21] uses the dictionary-based approach to build a lexicon for sentiment classification and uses three online dictionaries rich in vocabulary to collect thesauruses based on the seed words to improved reliability of the lexicon. Thesauruses are a collection of antonyms and synonyms to expand the lexicon more vocabulary.

Because he focused only on lexicon building, the result was slightly increased.

5   Big Data

Massive amounts of data are stored in communication sites. These data are increasing rapidly every year. Most of the data stored on the Internet has been produced in recent years. But most of the data produced may be useful when processed. Business, political, and social sites are the main source of data growth and the largest data incubator.

Big data can be defined as a collection and storage of huge amounts of information for final processing; there is a concept that gained momentum among the public in the early 2000s almost by Doug Laney which explain the current mainstream definition of big data. *Big data* is of *high volume*, *high* velocity, and *high* variety. High volume means large amounts of information are collected from different sources such as commerce, social media, and other communication channels and then stored in Hadoop support. High velocity is a high capability of retrieval data, sharing into slaves, process speed should be very high, high variety means all varieties of data processing, such as structured, unstructured, and semi structured.

Big data involves various tools including Hadoop, data science, MongoDB, data mining, Teradata, and Python. Hadoop is a big data framework to store and process high volumes of data with very high speed. Hadoop is an open source. Hadoop satisfies all characteristics of big data. So you can say Hadoop as a big data framework. Hadoop framework includes two main modules: Hadoop MapReduce and Hadoop Distributed File System.

- MapReduce is a processing technique and a program model for distributed computing based on Java. MapReduce algorithm involves two important tasks those are map and reduce. The map is A set of data which is converted to another set of

Fig. 4 Data science terms

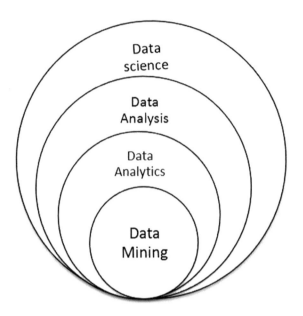

data (map is input value) Reduce job takes the output from a map; reduce is output of data value, the reduce task is always performed after the map job.

- Hadoop Distributed File System (HDFS) is designed to store a huge data sites reliably. HDFS creates several replicas of a huge data for reliability and then put them on compute nodes around the cluster.

Data Science: It is very suitable for data analysis, so it can give a boost of improvements to sentiment analysis. Data science is dealing with each structured data, unstructured data, and semi-structured data. In simple terms, it incorporates statistics, mathematical, analysis, signal processing, natural language processing, etc. Data science is an umbrella which involves several terms as illustrated in Fig. 4.

5.1 Big Data Tools in Sentiment Analysis

Using the modern program in big data is very important for sentiment analysis at the moment. The use of big data for sentiment is very appropriate and still at the beginning of growth. Most of the comments available in social media are unstructured. So, we can use Hadoop technology because Hadoop can deal with structured data, unstructured data, and semi-structured data. A research study is conducted to determine diabetes awareness among different segments of the population using Hadoop MapReduce, [21]. The author Monu Kumar [28] extracts and collect the data from social networks and analyze it by using big data technique. Hadoop MapReduce is processing technique of big data effectively, so priorities use Hadoop for data collection and analysis. Shirahatti [21] collects comments and reviews from Twitter using streaming tool flume then processes it in Hadoop MapReduce. The result of processing time taken was very less compared to others previous methods.

6   Sentiment Analysis Challenges

The researcher in this field faces several of constraints and challenges which come in the form of sentences or words vague difficult to identify. These constraints constitute an obstacle to analyze targeted data and may lead to an unreliable outcome. There are different types of these obstacles that pose challenging for sentiment analysis using one of the known questionnaires, simple questionnaires, or role-based questionnaires [22]. But the clear events is highly accepted and a candidate to obtaining good comments and high-quality. We will illustrate some types of challenges which are listed below.

i. **Success or fail from one side:**In many events that occur daily, expressions and phrases are often used to describe the win or loss of one side. For example, Brazil defeated Argentina 2-1; The Supreme Court ruled in favor of early marriage, the regime's army suppressed a popular uprising. In this case, if the person supports Brazil, early marriage, and the ruling regime, will consider these events as positive. but the person supporting Argentina and marriage over the age of eighteen, and the popular uprising will consider these events as negative. Also, we note that the transfer of the proceedings of the event as the win of one party or (as the loss of another party) does not mean that the speaker expresses a negative or positive opinion towards one of the parties mentioned. For example, when Sevilla defeated Real Madrid 2-1 in the Spanish League 2017, the news around the world was reported approximately that Real Madrid lost from Sevilla, instead of broadcasting the news Sevilla beat on Real Madrid. This does not refer that speakers or those involved in the broadcast reports expressed a negative opinion on the Real Madrid team, but the royal team was their focus and they thought that Real Madrid would not easily lose and owns with him forty matches without losing pre-Sevilla match.

ii. **Precisely understanding what is meant by opinion:** Sometimes, it is difficult for a person to understand the exact meaning of some sentences. For example, glad to expose the disciplinary committee steroids of player Noor. From this context, the target of the view seems unclear, do he mean Noor, Noor doping, or Noor doping that which are revealed. In this case, the person can conclude that the news carrier has a negative idea toward Noor doping and probably Noor in general.

iii. **Neutral reports of events**: When the announcer starts broadcasting the reports, the speaker must give a signal to his or her emotional state before describing the events or situations, because It is unclear whether these reports should be considered as emotional about evolution or presupposes whether the speaker is in a negative emotional state (Happy, angry, cheerful, sad,).

iv. **Detection abusive opinions and counterfeit reviews:** On the Web, there is a lot of spam information which contains spam and abusive reviews for sentiment classification; it is unacceptable to process data with a presence of fake data because this reduces the reliability of the results; we should initially identify unwanted messages and remove them and then the processing, These steps we can do through reviewer [29].

v. **The focus on one domain:** This challenge is a major obstacle to sentiment analysis because it depends mainly on a limited nature from sentiment analysis word; this may lead to focus on only one topic. For example, we may find in one domain several features and good performance; at the same time, these features may be very bad in some other domains [30].

vi. **Difficult acquisition of opinion mining software**: The software of opinion mining is very expensive, and their prices are high and can only be bought by governments and large organizations for now. These high prices exceed people's expectations and will remain a major obstacle to people wanting to get this software; the average person cannot buy this expensive software. Therefore, this software should be available to all categories of society without exception so that everyone benefits from them.

7 Conclusion

This study mainly focuses on the overview of different techniques used in the field of SA. Among thirty-eight papers recently published till 2017, we discussed the importance of opinions and comments on Web sites and how to extract them through certain techniques. We have noted important techniques in this area including the most famous as Naive Bayes and SVM are the more commonly utilized in machine learning algorithms, to solve sentiment classification problem. Many present researchers have improved the scope of this field. The aim of this study gives overview on these improvements and summarizes categories of articles given according to different sentiment analyses. The contribution of this study takes a brief look at the use of big data (Hadoop tools) in sentiment analysis field. Use a big data is a fairly new study in SA field. Therefore, our future work will focus on using big data techniques (Hadoop) in sentiment analysis field to give more effective and accurate results. This paper will be useful for new researchers and who has a desire to join this field. In this study, we covered all the techniques and the most famous in one paper and illustrated different techniques in (SA) approach for extracting and analytics sentiments associated with the polarity of positive or negative, or neutral on the topic selected.

References

1. Nasukawa Y (2003) Sentiment analysis: capturing favorability using natural language processing, IBM Almaden Research Center, CA 95120, https://doi.org/10.1145/945645.945658
2. Mohey D (2016) A survey on sentiment analysis challenges. J King Saud Univ Eng https://doi.org/10.1016/j.jksues.2016.04.002
3. Alessia D (2015) Approaches, tools and applications for sentiment analysis implementation. Int J Comput Appl 125(3)
4. Xu W , Ritter A, Grishman R (2013) Gathering and generating paraphrases from twitter with application to normalization
5. Hazra TK (2015) Mitigating the adversities of social media through real time tweet extraction system, IEEE, https://doi.org/10.1109/iemcon.2015.7344483
6. Semih Y (2014) Tagging accuracy analysis on part-of-speech taggers. J Comput Commun 2:157–162, https://doi.org/10.4236/jcc.2014.24021
7. El-Din DM (2015) Online paper review analysis. Int J Adv Comput Sci Appl 6(9)
8. Kaushik L (2013) Sentiment extraction from natural audio streams, IEEE https://doi.org/10.1109/icassp.2013.6639321
9. Vaghela VB (2016) Analysis of various sentiment classification techniques. Int J Comput Appl 140(3)
10. BiltawiL M (2016) Sentiment classification techniques for Arabic language a survey, IEEE, https://doi.org/10.1109/iacs.2016.7476075
11. Goel A (2016) Real time sentiment analysis of tweets using naive bayes, IEEE, https://doi.org/10.1109/ngct.2016.7877424
12. Hu M, Liu B (2004) Mining and summarizing customer reviews, seattle, Washington, USA, https://doi.org/10.1145/1014052.1014073
13. Kim S-M (2004) Determining the sentiment of opinions, ACM Digital Library, https://doi.org/10.3115/1220355.1220555
14. Mohammad S (2009) Generating high-coverage semantic orientation lexicons from overtly marked words and a thesaurus. In: Conference on empirical methods in natural language processing, pp 599–608
15. Miller GA (1993) Introduction to word net: an on-line lexical database
16. Hatzivassiloglou V, McKeown R (1998) Predicting the semantic orientation of adjectives, New York, N.Y.10027, USA
17. Medhat W (2014) Sentiment analysis algorithms and applications a survey. Ain Shams Eng J (Elsevier B.V.), 5(4):1093–1113
18. Soo-Min Kim, Determining the Sentiment of Opinions, International Journal, doi=10.1.1.68.1034, (2004)
19. Pang B, Lee L (2008) Opinion mining and sentiment analysis. https://doi.org/10.1561/1500000011
20. Niu Y (2005) Analysis of polarity information in medical text, PMC Jurnal
21. Park S (2016) Building thesaurus lexicon using dictionary based approach for sentiment classification, IEEE, https://doi.org/10.1109/sera.2016.7516126
22. Ramsingh J (2016) Data analytic on diabetic awareness with Hadoop streaming using map reduce in Python, IEEE, https://doi.org/10.1109/icaca.2016.7887979
23. Kim S-M, Hovy E (2006) Automatic identification of pro and con reasons in online reviews, ACM Digital Library
24. Trupthi M (2017) Sentiment analysis on twitter using streaming API, IEEE, https://doi.org/10.1109/iacc.2017.0186
25. Cambria E, Hussain A (2015) Group Using Lexicon Based Approach. Springer J https://doi.org/10.1007/978-3-319-23654-4
26. Akter S (2016) Sentiment analysis on Facebook group using lexicon based approach, IEEE, https://doi.org/10.1109/ceeict.2016.7873080
27. Yoshizawa A (2016) Machine-learning approach to analysis of driving simulation data, IEEE, https://doi.org/10.1109/icci-cc.2016.7862067

28. Istiaq Ahsan MN (2016) An ensemble approach to detect review spam using hybrid machine learning technique, IEEE, https://doi.org/10.1109/iccitechn.2016.7860229
29. Kumar M (2016) Analyzing Twitter sentiments through big data, IEEE, https://doi.org/10.1109/sysmart.2016.7894530
30. Abhinandan P, Shirahatti (2015) Sentiment analysis on Twitter data using Hadoop. Int J Eng Res Gen Sci 3(6)

An Efficient Automatic Brain Tumor Classification Using LBP Features and SVM-Based Classifier

Kancherla Deepika, Jyostna Devi Bodapati and Ravu Krishna Srihitha

Abstract Brain tumor detection is a tedious task which involves a lot of time and expertise. With each passing year, the world has always witnessed an increase in the number of cases of brain tumor. It is thereby apparent; that it is becoming difficult for the doctors to detect tumors in MRI scans, not only because of the increase in numbers but also, because of the complexity of the cases. So, the research in this domain is still ongoing as the world is in search of an exemplary and flawless method for an automated brain tumor detection technique. In this paper, we introduced a novel architecture for brain tumor detection which detects whether the given MR image is malignant or benign. Preprocessing, segmentation, dimension reduction, and classification are the major phases of our proposed architecture. On the MR images, T2-weighted preprocessing is applied to convert into grayscale images. In the next stage, features are extracted from the preprocessed images by applying local binary pattern (LBP) technique. Principal component analysis (PCA) is used to discard uncorrelated features. This reduced feature set is fed to the support vector machine (SVM) classifier to predict whether the given MR image is normal (benign) or abnormal (malignant). Experimental results on benchmark MR image datasets exhibit that the proposed method gives promising accuracy when compared to the existing work though it is simple.

Keywords Brain tumor · LBP · PCA · SVM

1 Introduction

Brain tumor is the anomalous development of cells in the human brain that disrupt the normal functioning of the brain. The tumor can be primary or metastatic depending on where the tumor cells originate from. They can even be classified as benign or malignant taking into consideration their cancerous nature. The images obtained

K. Deepika · J. D. Bodapati (✉) · R. K. Srihitha
Vignan's Foundation for Science, Technology and Research, Guntur, AP, India
e-mail: jyostna.bodapati82@gmail.com

© Springer Nature Singapore Pte Ltd. 2019
N. Chaki et al. (eds.), *Proceedings of International Conference on Computational Intelligence and Data Engineering*, Lecture Notes on Data Engineering and Communications Technologies 28, https://doi.org/10.1007/978-981-13-6459-4_17

through magnetic resonance imaging (MRI) are used for this purpose. The MRI scans can be taken from various planes such as axial, sagittal, and coronal. The MR images captured from the axial angle are best suited for brain tumor detection. Identifying whether a given MR image is normal (absence of tumor) or abnormal (presence of tumor) in complex cases is an arduous task.

With the advances in machine learning algorithms, many researches tried to address the issue of automatic tumor detection at the early stages. In the recent past, lot of research has been happening on this problem. Lack of abundant abnormal images is the major challenge faced in this automatic tumor detection.

Most of the existing works proposed in [1–4] follow several stages in their architecture. Commonly used stages in automatic tumor detection architectures are: preprocessing, segmentation, feature extraction, and classification.

Shil [1] proposed an improved brain tumor detection which follows all the above-mentioned algorithms and uses discrete wavelet transform (DWT) to extract features. The resultant high-dimensional data is the major problem with this method.

Islam et al. [2] proposed an approach to classify MR images of brain using BWT-KSVM, a hybrid approach. But the accuracy obtained by this method is not impressive compared to the other methods.

Mehdi Jafari [3] proposed an automatic brain tumor classification method that uses seeded region growing segmentation along with the neural network-based classification. Zhang et al. [4] proposed another naive approach to classify MR images of the brain that is making use of an adaptive chaotic PSO. In [3, 4] though the number of stages in their architecture are reduced to two, due to the use of feed-forward neural network (FFNN)-based classification this method cannot exhibit good performance with limited data. FFNN-based classifier's performance is dictated by the amount of training examples available. As we have already mentioned, availability of sufficient tumor data is the major problem.

Our proposed architecture addresses the issue of high-dimensional data by using LBP feature extraction and further applies PCA to further reduce the number of features required to represent data. Our proposed method reduces the complexity of the model by reducing the number of stages in the architecture. Feature extraction, dimension reduction, and SVM-based classification are the stages we used to keep the model simple. Our proposed method exhibited high performance on the benchmark datasets to classify the brain MRI images into cancerous or non-cancerous.

The rest of this paper is structured as follows. Section 2 introduces the associated work on automatic tumor detection. Section 3 describes the issues with the existing approaches and how the proposed method addresses those issues and also briefs the strengths of our methodology. Experimental results are reported in Sect. 4 along with a detailed comparative study to show the effectiveness of the proposed work followed by conclusions and future work.

2 Related Work

Brain tumor is the anomalous development of cells in the human brain that disrupt the normal functioning of the brain. The tumor can be primary or metastatic depending on where the tumor cells originate from. They can even be classified as benign or malignant taking into consideration their cancerous nature. The images obtained through magnetic resonance imaging (MRI) are used for this purpose. The MRI scans can be taken from various planes such as axial, sagittal, and coronal. The MR images captured from the axial angle are best suited for brain tumor detection. Identifying whether a given MR image is normal (absence of tumor) or abnormal (presence of tumor) in complex cases is an arduous task.

Shil [1] proposed an advanced brain tumor detection and classification methodology. Their proposed scheme comprises of several number of stages. Initially, the images are converted to grayscale and Otsu Binarization is applied on the grayscale images. Then K-means clustering is applied to segment the affected part from the rest of the image. From the segmented image, features are extracted using discrete wavelet transform (DWT) which is followed by PCA-based dimensionality reduction. In the final stage, support vector machine (SVM)-based classifier is used to classify whether the given MR image is malignant or benign. They could achieve an accuracy of 99.33%.

Islam et al. [2] proposed a novel hybrid approach called BWT-KSVM for classifying the MRI images as normal or abnormal. The images are converted to grayscale, the noise is dismissed using median filter and the contrast of the images is enhanced. Using Otsu thresholding and K-means clustering, the segmentation is done. The feature vector comprises of the spatial features extracted using GLCM and BWT. Feature reduction is performed for the transformed features using PCA. SVM is used as the classification algorithm and an accuracy of 95.2% is obtained.

Mehdi Jafari [3] proposed an architecture that uses seeded region growing segmentation and neural network-based classification. PCA is applied on extracted DWT features to further reduce the dimensions. To classify the images, a supervised feed-forward neural network-based classifier is used. The network uses the standard back-propagation algorithm to learn the parameters of the network. Though complex, this technique achieves 99.8% of accuracy.

Zhang et al. [4] proposed another naive approach to classify MR images of the brain that is making use of an adaptive chaotic PSO. In this method, DWT is applied initially to extract features from images and then PCA is applied for dimensionality reduction. The reduced features are passed at the input layer of the feed-forward neural network (FFNN). The parameters of network are further optimized via adaptive chaotic particle swarm optimization (ACPSO). To enhance generalization, a variant of cross validation called stratified K-fold cross validation is used. With this method, an accuracy of 98.75% is achieved for classifying the brain MRI images into abnormal and normal.

The proposed study aims at classifying the brain MRI scans into normal or abnormal MR images with high accuracy. Our proposed architecture is simple in comparison with the existing works in terms of the number of stages used in the model. Feature extraction, dimension reduction, and SVM-based classification are the stages we followed to keep the model simpler than existing models. Though our proposed method is simple it exhibits 100% accuracy on the bench mark datasets to classify the brain MRI images into benign or malignant.

3 Proposed Method

The objective of the reported methodology is given an MR image classify it as normal or abnormal. The MR images are taken as input and preprocessed. Later LBP-based features are extracted from these images and feature reduction is applied to remove uncorrelated features. These features are used in the next phase to train the SVM-based classifier that predicts the normality or abnormality of an input MR image. Figure 1 describes the architecture of the proposed work.

A. *Preprocessing*

Images are preprocessed to transform and enhance them in order to best suit for the task that is under consideration. In the preprocessing stage of our proposed model, the input brain MR images are transformed to gray-level images. Any detail of the MR image that is trivial for the task being performed is treated as noise. For the detection of brain tumor, considering color information is futile. Thus, the images are converted to grayscale.

B. *Feature Extraction*

Feature extraction is the process of extracting features from the data that are intended to be informative and facilitate subsequent learning. Local binary patterns (LBP)

Fig. 1 Proposed architecture

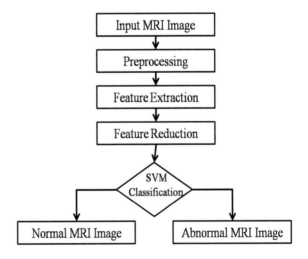

[7, 8] is one of the most popular texture-based features that is used for classification in computer vision. It is a statistical and texture-based method in which a histogram is computed by comparing a pixel value in a cell with its neighboring pixels. The histogram is then normalized and all of them are concatenated. The length of the histogram depends on the number of neighbors taken into consideration.

The employment of uniform patterns reduces the length of the feature vector. Uniform patterns are used to implement a simple rotation invariant descriptor. A binary pattern is considered uniform if it contains at most two 0-1 or 1-0 transitions. The histogram has a separate bin for each uniform pattern and one common bin for all non-uniform patterns. The use of uniform patterns reduces the length of the feature vector for a single cell from 256 to 59. Literature shows that LBP along with HOG gives improved classification performance.

C. *Discard Uncorrelated Features*

All the features extracted may not prove to be useful for the task being performed. Dimensionality reduction is the process of reducing the number of features used to represent the data. The purpose of dimensionality reduction is discarding the features that are uncorrelated while retaining the most correlated features.

To keep our model simple, the extracted LBP features are projected in a dimensional space using a linear feature reduction method called PCA. PCA linearly maps the data to a low dimensional space so as to maximize the variance of data in that representation. The features in the projected lower-dimensional space are uncorrelated. The obtained principal components account for the maximum variability in data.

D. *Classification*

SVM is the popular learning algorithm which classifies the given data with a maximum margin separating hyperplane. SVM follows the principle of structural loss (SRM) instead of the empirical loss as used by many other classifiers. By using SRM, it would be possible to attain a simple model rather than a complex model which works better on training data alone. According to the literature, SVM generalization is better than other models even if the model is trained with limited training data. Hence, we have selected SVM based on classifier in our proposed model.

The data projected using PCA is fed to SVM-based classifier to predict the normality or abnormality of a brain MR image. This SVM-based classifier projects the data onto a higher-dimensional space using kernel functions where the data can be linearly separable in the higher-dimensional space. The objective of SVM is to find the maximum margin hyperplane in the projected space. The separating hyperplane with maximum margin can be defined as:

$$wx + b = 0 \qquad (1)$$

where w is a weight vector and b is a scalar, often referred to as a bias.

The problem is to determine a hyperplane such that it satisfies Eq. (2) for all the positive class examples and satisfies Eq. (3) for all the examples belonging to the negative class.

$$w.x + b \geq 1 \tag{2}$$

$$wx + b \leq -1 \tag{3}$$

This problem can be posed as a minimization problem and can be solved using quadratic programming solver. Boundary function can be defined as:

$$d(x) = \sum_{i=1}^{l} y_i \alpha_i K\left(x_i x^T\right) + b_0 \tag{4}$$

where y_i represents class label, α_i and b_0 are Lagrangian co-efficient and bias, respectively. l is the number of support vectors. Given any test sample, x, if the above boundary function evaluates a positive value, it indicates that the sample belongs to positive class and it belongs to negative class otherwise.

In Eq. (4), $K\left(X_i X^T\right)$ is the kernel function applied on the pair of data X_i and X^T. If the data has to be transformed into a higher dimension so as to classify better, it can be achieved by obtaining the inner product of the data in the higher dimension rather than computing the exact transformation of the data. Computing the inner product is far easier than computing the actual points in the higher dimension. This can be achieved with the help of the kernel trick.

$$K\left(X_i, X_j\right) = \varphi(X_i).\varphi\left(X_j\right)$$

Following are the popular kernels that are used in the literature to achieve better available to achieve this. They are:

Linear kernel: $K\left(X_i, X_j\right) = \left(X_i^T.X_j\right)$
Polynomial kernel: $K\left(X_i, X_j\right) = \left(X_i.X_j + 1\right)^d$
Radial basis function kernel: $K\left(X_i, X_j\right) = e^{-||X_i - X_j||^2/2\sigma^2}$
Sigmoid kernel: $K\left(X_i, X_j\right) = \tanh\left(kX_i.X_j - \delta\right)$.

4   Experimental Results

The objective of our proposed work is to efficiently classify the given MR images. To exhibit the efficiency of our proposed method, we have used benchmark dataset namely Brats15 and Midas: Healthy human brain database. This dataset contains 100 normal and 180 abnormal images, a total of 280 MR images. The abnormal images contain both high grade as well as low grade glioma scans. These are gathered from

Table 1 Composition of dataset

Dataset	Train set	Test set
Normal	20	80
Abnormal	40	140

Table 2 Confusion matrix

	Normal	Abnormal
Normal	80	0
Abnormal	0	140

the Brats15 dataset. The normal images are obtained from the Midas database. 20 out of the 100 normal MR scans are used for training and the rest are used for testing. Out of the 180 abnormal MR scans, 40 images are used for training and 140 images are used for testing. The images are in mha format, and VV tool is used to visualize the images. T2-weighted images were collected for performing this experiment (Table 1).

After training the SVM with the training dataset, it is tested with some other MR scans and the performance is evaluated using some metrics such as accuracy, sensitivity, and specificity (Table 2).

$$\text{Accuracy} = \frac{TP + TN}{P + N} = \frac{80 + 140}{80 + 140} = 100\%$$

$$\text{Sensitivity} = \frac{TP}{P} = \frac{80}{80} = 100\%$$

$$\text{Specificity} = \frac{TN}{N} = \frac{140}{140} = 100\%$$

The proposed scheme has achieved an accuracy, sensitivity, and specificity of 100%. By reducing the complex stages of the architecture, we could improve the performance of the classifier.

5 Conclusion and Future Work

The proposed system gives exceptional results with a simple architecture. Neither a complex feature extraction technique nor a complex classification technique was used. With a very simple design, the proposed system achieved high accuracy.

In the future, we aim to work on new models for automating glioma detection. It is a very challenging task and also an important one to work on because the manual detection of glioma involves many tests and takes a lot of time too. In order to reduce all of this hassle, we aim to propose effective models for glioma detection.

References

1. Shil SK et al (2017) An improved brain tumor detection and classification mechanism. In: Proceedings of IEEE conference on information and communication technology convergence (ICTC), 2017
2. Zhang et al (2010) A novel method for magnetic resonance brain image classification based on adaptive chaotic PSO. Prog Electromagn Res 109:325–343
3. Islam A et al (2017) A new hybrid approach for brain tumor classification using BWT-KSVM. In: Proceedings of advances in electrical engineering, 2017
4. Bodapati JD et al (2010) A novel face recognition system based on combining eigenfaces with fisher faces using wavelets. Proc Comput Sci 2:44–51
5. Bodapati JD et al (2014) Scene classification using support vector machines with LDA. J Theoretical Appl Inf Technol 63(3)
6. Bodapati JD et al (2010) An intelligent authentication system using wavelet fusion of K-PCA, R-LDA. In: IEEE international conference on communication control and computing technologies (ICCCCT), 2010, pp 437–441
7. Pradhan D (2017) Enhancing LBP Features for Object Recognition using Spatial Pyramid Kernel. Int J of Comput Math Sci 6(6):105–109
8. Harris S et al (2017) LBP features for hand-held ground penetrating radar. In: Detection and sensing of mines, explosive objects, and obscured targets XXII, vol 10182. International Society for Optics and Photonics, 2017
9. Bullitt E, Zeng D, Gerig G, Aylward S, Joshi S, Smith JK, Lin W, Ewend MG (2005) Vessel tortuosity and brain tumor malignancy: a blinded study. Acad Radiol 12:1232–1240
10. Menze BH, Jakab A, Bauer S, Kalpathy-Cramer J, Farahani K, Kirby J, Burren Y, Porz N, Slotboom J, Wiest R, Lanczi L, Gerstner E, Weber MA, Arbel T, Avants BB, Ayache N, Buendia P, Collins DL, Cordier N, Corso JJ, Criminisi A, Das T, Delingette H, Demiralp Γ, Durst CR, Dojat M, Doyle S, Festa J, Forbes F, Geremia E, Glocker B, Golland P, Guo X, Hamamci A, Iftekharuddin KM, Jena R, John NM, Konukoglu E, Lashkari D, Mariz JA, Meier R, Pereira S, Precup D, Price SJ, Raviv TR, Reza SM, Ryan M, Sarikaya D, Schwartz L, Shin HC, Shotton J, Silva CA, Sousa N, Subbanna NK, Szekely G, Taylor TJ, Thomas OM, Tustison NJ, Unal G, Vasseur F, Wintermark M, Ye DH, Zhao L, Zhao B, Zikic D, Prastawa M, Reyes M, Van Leemput K (2015) The multimodal brain tumor image segmentation benchmark (BRATS). IEEE Trans Med Imaging 34(10):1993–2024. https://doi.org/10.1109/TMI.2014.2377694
11. Bakas S, Akbari H, Sotiras A, Bilello M, Rozycki M, Kirby JS, Freymann JB, Farahani K, Davatzikos C (2017) Advancing the cancer genome atlas glioma MRI collections with expert segmentation labels and radiomic features. Nat Sci Data 4:170117. https://doi.org/10.1038/sdata.2017.117

Analyzing Student Performance in Engineering Placement Using Data Mining

Krishnanshu Agarwal, Ekansh Maheshwari, Chandrima Roy, Manjusha Pandey and Siddharth Swarup Rautray

Abstract Data mining is the practice of mining valuable information from huge data sets. Data mining allows the users to have perceptions of the data and make convenient decisions out of the information extracted from databases. The purpose of the engineering colleges is to offer greater chances to its students. Education data mining (EDM) is a process for analyzing the student's performance based on numerous constraints to predict and evaluate whether a student will be placed or not in the campus placement. The idea of predicting the performance of higher education students can help various institutions in improving the quality of education, identifying the pupil's risk, upgrading the overall accomplishments, and thereby refining the education resource management for better placement opportunities for students. This research proposes a placement prediction model which predicts the chance of an undergrad student getting a job in the placement drive. This self-analysis will assist in identifying the patterns, where a comparative study between two individual methods has been made in order to predict the student's success and a database has been generated.

Keywords Data mining · Campus placement prediction · Classification · KNN · Random forest

K. Agarwal (✉) · E. Maheshwari · C. Roy · M. Pandey · S. S. Rautray
School of Computer Engineering, Kalinga Institute of Industrial Technology (KIIT)
Deemed to be University, Bhubaneswar, Orissa, India
e-mail: agarwalkrishnanshu@gmail.com

E. Maheshwari
e-mail: ekansh031998@gmail.com

C. Roy
e-mail: chandrima.roy.1914@gmail.com

M. Pandey
e-mail: manjushafcs@kiit.ac.in

S. S. Rautray
e-mail: siddharthfcs@kiit.ac.in

© Springer Nature Singapore Pte Ltd. 2019
N. Chaki et al. (eds.), *Proceedings of International Conference on Computational Intelligence and Data Engineering*, Lecture Notes on Data Engineering and Communications Technologies 28, https://doi.org/10.1007/978-981-13-6459-4_18

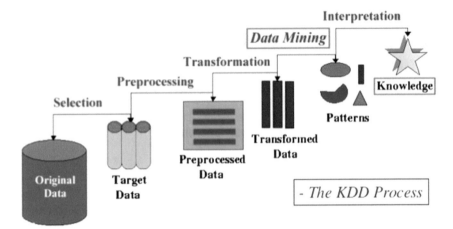

Fig. 1 Knowledge data discovery process

1 Introduction

Placement is considered the main factor for measuring the success of a student. The academic performance of a student depends on multiple factors and analyzing those factors is quite challenging. In recent years, there has been a drastic rise in people's interest rate in the practice of data mining for educational and academic purposes. Research in education is an upcoming area which involves data mining and has very different requirements from other fields for the assessment of the student's performance. The objective is to do a relative study of student's performance through diverse algorithms taking into consideration various factors which are influential in a student's success in placements [1].

"The education system is deteriorating slowly day by day," is a rising thought upon which the people across the world agree nowadays. By analyzing certain factors, the educational institutions can refine models and make them full of insights and apparition. The mentors, the parents, and the education system as a whole will be benefited as they will know the areas that needs lateral thinking and can get along their academic performance which will help them to build a better academics and thereby an experienced carrier.

Knowledge discovery in databases (KDD) [2] is the process of finding useful information from a dataset represent in Fig. 1. It includes multidisciplinary activities. The knowledge discovery phase involves various steps like data cleaning, data integration, selection, transformation, mining, and finally pattern evaluation. Initially, we need to eliminate the noise and inappropriate data. This process of removal or reduction of the unwanted information is known as data cleaning.

Secondly, multiple data sources are combined in order to promote versatile data and information. This whole technique comes under data integration. Once the data is cleaned and integrated, we proceed toward the most important step which is selection of data. This is done according to the needs and requirements. According to the requirement of the analysis, the relevant data are retrieved from the database. Once the desired data are fetched [3], we transform the data and consolidate it into appropriate forms for undergoing the mining process. Finally, the aggregation process is operated and performed. There are various mining methods. To extract valuable data patterns, we choose one of the techniques and apply on the data. On evaluation of the data patterns, we drive to conclusions useful for our research. The in-gained knowledge is finally communicated in the form of graphs, tables, statistics, etc. This helps us in various analysis and research [4]. This process has reached its optimum in the past decade. The recognized pattern in the data possesses varying degree of certainty. It contains various approaches to discovery that include inductive learning, Bayesian statistics, knowledge acquisition, etc. The main goal is mining of high-level data from low-level knowledge.

1.1 Classification Techniques

Classification is the process of identifying an entity to which set or class of category it belongs to, based on the training dataset containing observations whose class in known [5]. There are various techniques of classification known to us like linear classifiers which include logistic regression and naïve Bayes classifier, support vector machines, decision trees, and many more. In this paper, we have used two techniques K-nearest neighbor and random forest. The K-nearest neighbor is a supervised classification algorithm. It uses labeled points to learn to label other points. For labeling a new point, it finds the nearest points to that point and checks the labels of those nearest points. The label of most of the nearest points is the label of that new point, where k is the number of neighboring points it checks.

Random forest is another supervised learning algorithm which works by creating a group of random uncorrelated decision trees that are called forest. It then sums up the votes from various decision trees to decide the best final class for the test data.

The paper can has been divided into sections whose description has been given here. In Sect. 2 the papers refereed for the papers are discussed. In Sect. 3, what is the need for the proposal has been discussed. In Sect. 4, the description of data and how data were collected are given with the proposed mechanism for the flow of data. In Sect. 5, the experimental setup has been discussed followed by Sect. 6 in which the result and analysis part of the paper has been discussed. In Sect. 7, the conclusion of paper has been discussed.

2 State-of-the-Art

Enormous amount of work has been done in the current years to provide a comprehensive review of data mining in education. Data mining represents potential areas

Table 1 Study and getting idea from different papers

Paper and author name	Objective	Algorithm used
"A data mining techniques for campus placement prediction in higher education" published by Tansen Patel et al. in the year 2017 [2]	Implement data mining technique to enhance prediction for campus placement	K means, farthest first, filtered cluster, hierarchical cluster, and make density-based clustering
"Student prediction system for placement training using fuzzy inference system" by Ravi Kumar Rathore et al. in the year 2017 [7]	Identify and improve the performance of the students	Fuzzy inference system
"Predicting student performance: a statistical and data mining approach" by V. Ramesh et al. in the year 2013 [6]	Prediction of student's performance	Multilayer perceptron (MLP) Naïve Bayes, multilayer perception, SMO, J48, and REP tree algorithms
"A placement prediction system using k-nearest neighbors classifier" published by Animesh Giri et al. in the year 2016 [8]	Predicts the probability of a student getting placed in an IT company	K-nearest neighbors, logistic regression, and support vector machine
"PPS—placement prediction system using logistic regression" by Ajay Shiv Sharma et al. in the year 2014 [9]	Predict the placement of a student	Logistic regression and gradient descent
"Educational data mining for student placement prediction using machine learning algorithms" by K. Sreenivasa Rao et al. in the year 2018 [10]	Predict whether a student will get placed or not in the campus placement	J48, Naïve Bayes, random forest, and random tree

of researches in the education sector. It contains specific requirements which other fields lack. Different methodologies and techniques of data mining have been used to predict students' success, using the collected data. Some of the vocation done by different researchers has been described below (Table 1):

3 Research Proposal

Placement is considered to be a very crucial for any undergraduate student. Students have a lot of perplexity about their weaknesses, strengths, and which type of job profile they are suitable for [6]. Such lack of knowledge leads to a lot of unsuitable placements of those students which results in unsatisfactory job profiles. This model can help such students to mark the best suitable job profiles for them and can help them with the necessities of those profiles as well. Another problem faced by the students is that they do not know on what factors have how much impact on their placement.

4   Proposed Framework

Student's learning outcomes in the form of various assessments are the indicator of a student's performance. These outcomes give us important information about a pupil's field of learning and the extent to which he/she is interested in those fields. However, the grading system is the standard method of measuring various levels of accomplishments. A grade is mainly divided into two types: grade point average (GPA) and cumulative grade point average (CGPA). CGPA is calculated by taking an average of the grades obtained by the student up to the last semesters he/she has appeared for.

4.1   Dataset Preparation

The dataset that is used in this paper is the placement statistics of Kalinga Institute of Industrial Technology (KIIT) Deemed to be University, Bhubaneswar, Orissa for the final year academic batches of B. Tech. For this, a survey using Google form was conducted. The data comprise of responses from 306 students.

4.2   Data Description

The data that are taken for this model are from the placement log of the college. The attributes that were found to be most significant for the placement of a student were taken into examination. Table 2 depicts the same.

Though there are numerous factors that can influence the selection of a student for a particular campus placement by a company, here in this paper we have taken into consideration some very basic but important features like CGPA, quantitative aptitude, coding languages known, English speaking skills, number of projects and internships done, etc.

Data preprocessing is a technique of data mining which involves transforming of the raw data into an understandable format. Separating data into training and testing datasets is an essential portion of assessing data mining models which is represent in Fig. 2. After separation of data into training and testing sets, the set having major part of data is used for training purpose while the set having minor part of data is used for testing purpose.

Table 2 Data variables considered for classification

Variable	Variable range	Variable type
CGPA	0–10	Numeric, continuous
Efficiency in quantitative aptitude	Bad, average, good, very good, and excellent	String
English written/spoken skill	Bad, average, good, very good, and excellent	String
Proficiency in coding languages	C, C++, JAVA, Python, and R	String
Backlogs	0–8	Numeric, discrete
Publications/research paper	0–8	Numeric, discrete
Number of projects	0–8	Numeric, discrete
Placed or not	Yes or no	String
Type of profile offered	Consultancy or technical	String

Fig. 2 Placement prediction system architecture

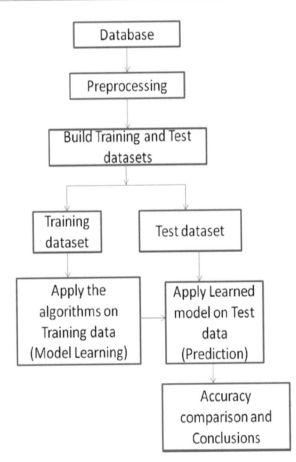

5 Experimental Setup

The hardware configurations of the desktop used are processor speed of 3.20 GHz, RAM of size 8.00 GB, and system type of 64 bit operating system with a ×64-based processor. The software configurations used for this introspection include Python 3.6 version running on top of Windows 10. Python is an open-source interpreted high-level programming language. It has a design philosophy that emphasizes code readability. Scikit-learn, a Python machine learning library was also used along with matplot, a plotting library. Package of Pandas has been used, which is a Python package and provides fast, flexible data structures which manipulate numerical tables as well as time series. An active Internet connection for data recovering over network was also required.

6 Result and Analysis

This analysis helped us to know the exact factors affecting a student's performance in campus placement. It gave a clear idea about which factors directly and indirectly lead to the increase in the overall accomplishments and what are the key elements that may affect adversely. We have compared two classification techniques namely: k-nearest neighbor and random forest algorithm.

Figure 3 shows the results obtained from KNN algorithm on the basis of CGPA, number of projects and coding languages known with an accuracy of 93.54%. Here, two classes have been taken into consideration namely placed and unplaced represented by blue and green colors, respectively. However, when random forest algorithm was applied to the same data, as shown in Fig. 4, 83.87% accuracy is achieved which is less than the accuracy achieved by the KNN classifier. Thus, both the algorithms resulted in the output that the number of project and proficiency in coding languages is directly proportional to student's CGPA.

Secondly, we also analyzed how English written, English spoken, and quantitative aptitude affect the progress of a student in placement activity with an accuracy of 80.54%. They can be categorized in numeric terms from 1 to 5 where 1 refers BAD, 2 = AVERAGE, 3 = GOOD, 4 = VERY GOOD and, EXCELLENT is denoted by 5. From Fig. 5, it can be implied that if a student is fluent in English, as well as the student has good English written skill and excellent in quantitative aptitude, then he/she may be placed. Whereas Fig. 6 shows the results obtained from random forest classification algorithm with 80.64% accuracy.

Figure 7 shows the results obtained from KNN algorithm on the basis of English written, English spoken, and quantitative aptitude. The accuracy of KNN classifier when test data is given as input is 90.32%. Whereas Fig. 8 represents the results obtained from random forest classification algorithm on the basis of number of research paper, projects, and coding languages known with 93.54% accuracy.

Fig. 3 KNN on the basis of
CGPA, number of projects,
and coding languages known

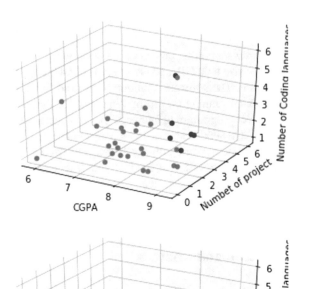

Fig. 4 Random forest on the
basis of CGPA, number of
projects, and coding
languages known

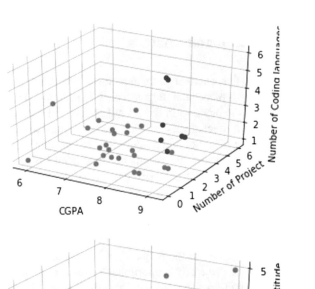

Fig. 5 KNN on the basis of
English written, English
spoken, and quantitative
aptitude

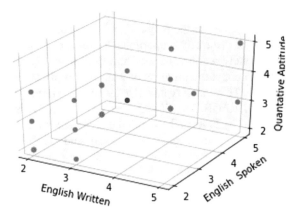

Fig. 6 Random forest on the basis of English written, English spoken, and quantitative aptitude

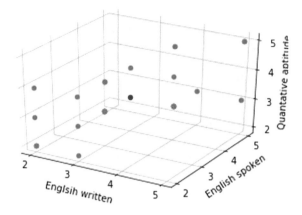

Fig. 7 KNN on the basis of number of research paper, projects, and coding languages known

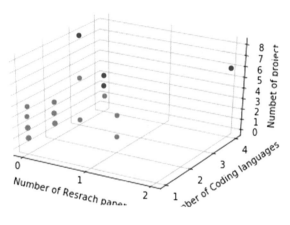

Fig. 8 Random forest on the basis of number of research paper, projects, and coding languages known

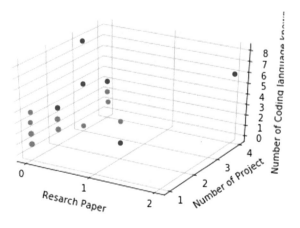

Figure 9 shows how the different factors are affecting a student's performance in placement activities. It can be seen that B. Tech CGPA attribute is contributing more toward placements of students. Average CGPA of placed student is calculated as 8.546 and average CGPA of unplaced student is 7.53. The impact of this factor on placement can be categorized in numeric terms from 0.0 to 0.5 where 0.0 refers to very low impact and very high impact is denoted by 0.5. While CGPA has an impact factor of 0.443 followed by number of projects with a factor of 0.135 and then by 0.133 the number of coding languages known. Comparing these with the results obtained from placed people by dividing the profile into technical based and consultancy based, it was found that CGPA impact factor goes down to 0.337 followed by the number of projects with a factor of 0.204 and then by English written with a factor of 0.093.

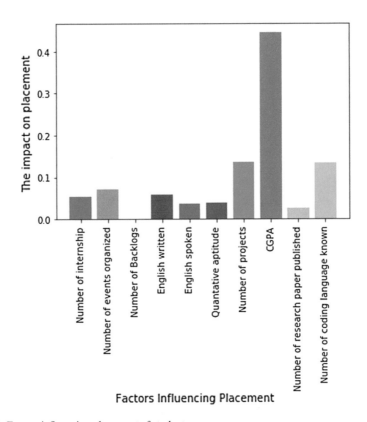

Fig. 9 Factors influencing placement of students

7   Conclusion

In every institution placement analysis is very crucial, which is beneficial for students seeking placement after completion of their course work. The experimental results of classification algorithms on student's placement datasets are performed here. The performance of the various classification algorithms is compared based on time taken to predict the classes. The results obtained by the application of different algorithms are tabulated and analyzed which shows that KNN algorithm gives an average accuracy of 88.13% while random forest gives us an average accuracy of 86.01%. The highest accuracy of both the algorithms was found to be 93.54%.

References

1. Parekh S et al (2016) Results and placement analysis and prediction using data mining and dashboard. Int J Comput Appl 137(13)
2. Patel T, Tamrakar A (2017) A data mining techniques for campus placement prediction in higher education. Indian J Sci Res 14(2):467–471
3. Roy C, Rautaray SS, Pandey M (2018) Big data optimization techniques: a survey. Int J Inf Eng Electr Bus 10(4):41
4. Roy C, Pandey M, Rautaray SS (2018) A proposal for optimization of horizontal scaling in big data environment. In: Kolhe M, Trivedi M, Tiwari S, Singh V (eds) Advances in data and information sciences. Lecture Notes in Networks and Systems, vol 38. Springer, Singapore
5. Adhatrao K (2013) et al Predicting students' performance using ID3 and C4. 5 classification algorithms. arXiv preprint arXiv:1310.2071
6. Ramesh V, Parkavi P, Ramar K (2013) Predicting student performance: a statistical and data mining approach. Int J Comput Appl 63(8):35–39
7. Rathore RK, Jayanthi J (2017) Student prediction system for placement training using fuzzy inference system. ICTACT J Soft Comput 7(3):1443–1446
8. Giri A et al A placement prediction system using k-nearest neighbors classifier. In: 2016 second international conference on cognitive computing and information processing (CCIP), IEEE, 2016
9. Sharma AS et al (2014) PPS—placement prediction system using logistic regression. MOOC. In: 2014 IEEE international conference on innovation and technology in education (MITE), IEEE, 2014
10. Sreenivasa Rao K, Swapna N, Praveen Kumar P (2018) Educational data mining for student placement prediction using machine learning algorithms. Int J Eng Technol 7(1.2):43–46. [Online]. Web 13 Aug 2018

RANK-Swapping Unblocked Row (RSR) Memory Controller

Arun S. Tigadi, Hansraj Guhilot and Pramod Naik

Abstract The main parts of the current-day memory system are a memory controller and a memory device. The task of the controller is to coordinate the requests from the CPU, DMA, and other devices. Command bus and data bus act as a bridge between two components. The front end of the memory controller does the work of generating commands related to the respective request. The timing issue and arbitration of these commands are taken care by the back end of the memory controller. The critical requirement of these memory controllers is to provide service to all the requestors without violating timing issues. The real-time multi-core systems (Bui et al in Temporal isolation on multiprocessing architectures. IEEE, 2011) [1] have many resources shared between the cores. This makes timing analysis harder in case of these systems. We use different methods to analyze these systems and make the assumption that "Access latency of single request" doesn't depend on different cores. The problem in deriving upper bounds is mainly due to the complex nature of multi-core systems [1]. These systems use DDR RAM as their main memory. These memories are partitioned into RANKS and BANKS which can support parallelism. Moreover, internal caching used by DRAM makes locality of references significant. The present memory controllers distribute memory request based on the command sequences generated, but in real time, this nature will not take advantage of locality. We will introduce a new design for a multi-core system for DDR devices. This employs RANK-swapping and open-row policies. The main advantage of this scheme is a significant reduction in the worst-case latency. This is mainly due to the creative RANK-switching mechanism. The memory is portioned into RANKS, BANKS, and rows to store an array of data. From this, we can isolate hard and soft requestors by assigning the same RANK to a respective type of requestors. This architectural

A. S. Tigadi (✉) · H. Guhilot
KLE Dr. M.S.S. CET, Belagavi, India
e-mail: arun.tigadi@gmail.com

H. Guhilot
e-mail: hansraj.g@gmail.com

P. Naik
CoreEL Technologies PVT LTD, Bengaluru, Karnataka, India
e-mail: pramodnaik40@gmail.com

© Springer Nature Singapore Pte Ltd. 2019
N. Chaki et al. (eds.), *Proceedings of International Conference on Computational Intelligence and Data Engineering*, Lecture Notes on Data Engineering and Communications Technologies 28, https://doi.org/10.1007/978-981-13-6459-4_19

improvisation improves the performance. Read–write latency can be taken out of the picture to improve bus utilization. By this technique, the latency for hard requestors is made predictable because it only depends upon a number of other requestors in the same RANK. Arbitration in hard requestor is mainly focused on latency in the worst case (Wu in Worst case analysis of DRAM latency in hard real-time systems, 2013) [2]. Arbitration for soft requestors is focused on throughput optimization.

Keywords BANK · RANK · Real-time · SDRAM · FPGA

1 Introduction

Generally, memory devices at present are created as RANKS, and these RANKS are further divided into BANKS. The buses can be accessed simultaneously unless there is no collision of requests. The storage cells are arranged in columns and rows. These cells make up BANKS; each BANK has a row buffer. The data in the DRAM row is accessed by issuing appropriate commands on command bus. In a single-channel system, one command and one data bus are present. First, the "activate," i.e., ACT, command is issued followed by CAS command for reading or writing.

The ACT command helps in loading data into a row buffer. After this, reading or writing operation is done. In the next request, if data from the different rows is needed, then the current row is "precharged"; i.e., the previously accessed row from the array is updated. At the end for maintaining data integrity, periodic "refresh," i.e., REF, command is issued.

This is done for all the BANKS and RANKS throughout the memory array. It takes one clock cycle for servicing each command. After CAS command, data movement on data bus happens in terms of bursts. The amount of data transferred is given by the product of burst length and bus width (BL × WBUS). In case of DDR devices, the data is sent on both the clock edges, so the time required for single transfer is given by BL/2 (Fig. 1).

At any moment, if a row is copied on the row buffer, it is known to be "open" and rest all rows are closed. As the name suggests, the "open requests" are the requests for accessing open row and similarly "close requests" are requests for accessing closed rows. It is the row buffer status which determines the generation of different commands when the request for memory access is given to front end. Only CAS command is needed for an open request since the required content will be available on the row. If we consider a request to a closed row, then we require PRE and ACT

Fig. 1 Memory controller

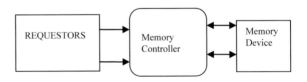

commands together in case of a different row. The memory controller will choose a particular policy from the different choices available depending on the importance of the situation. For an open-row policy, the row will be kept open, and for a close-row policy, it will be closed.

The job of the controller is to direct different requests to correct BANKS and RANKS. BANK interleaving is a technique which allows requestors to access all the BANKS and RANKS simultaneously. The main disadvantage of this method is that any row may get closed because of mutual interference caused by different requestors. To avoid this situation, private banking is used where different requestors have separate memory BANKS. The action of one requestor does not affect the other.

2   Related Contribution

The related works on memory controller design carried out by researchers can be classified under different implementation categories such as close-row, open-row, critical, mixed critical, RANK-switching, and arbitration policies. Let us analyze the related work under each of these categories. First, we will be looking into the related work focusing on real-time memory controllers with close-row policy. The work done on the Analyzable Memory Controller (AMC) [3], and the Predator employs close-row policy designed for the critical systems [4]. There will not be any guarantee that a row opened by a request will not be closed by the other requests. Here, the latencies can be quite larger than the open-row policy.

The work done by Yonghui et al. presents the architecture of a dynamically scheduled real-time memory controller. The paper analyzes to minimize the worst-case execution through close-page policy and BANK parallelism with interleaved BANK mapping. Further, this paper specifically addresses the issue of having either a fixed or variable transaction size for the real-time memory controllers. When it comes to the RANK-switching techniques, the research paper [5–7] uses the RANK-switching methods. Wang et al. proposed an algorithm to increase the BANK-level parallelism by scheduling a group of read or write commands to the same RANK. At that juncture, another set of CAS commands will be issued to the other BANKS.

3   Timing Constraints

The main task of the controller back end is to satisfy all the timing constraints. These include timing requirements between different commands. The JEDEC standard gives timing constraints for memory devices [8]. We will discuss these timing constraints using the following examples. These figures indicate different commands issued and data movement on the data bus. Different symbols indicate the different commands. The ACT, PRE, and R/W represent the active, precharge, and the

Fig. 2 Timing restrictions for the RANK containing same BANKS

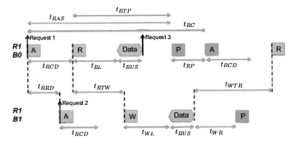

Fig. 3 Different RANKS and its timing constraints

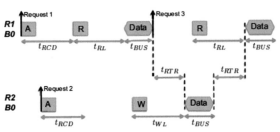

read/write operation. The horizontal line represents the timing constraints, and the vertical lines represent the time at which command is issued.

Figure 2 shows the timing constraints w.r.t. same RANK containing different BANKS. In this case, three close requests try to access the same RANK. The BANK 0 is being accessed by two requests 3 and 1, and the BANK 1 is being accessed by close request 2. From Fig. 2, it can be seen that the timing constraint t_{RTW} comes as the time delay which occurs between the request 1 read command and request 2 write command. Until this condition is met, we cannot issue a write command [9].

In the same manner, before issuing read command, timing requirement of t_{WTR} should be satisfied (Fig. 3).

From the above examples, we can make out certain points as listed below.

1. By comparing latencies of closed and open requests, it can be seen that the latency of open request is much lesser. It is due to different timing constraints associated with PRE and ACT.
2. Using a single RANK, one cannot take advantage of BANK parallelism. Switching between different BANKS in the same RANK involves tRRD, and switching between read and write commands involves tRTW and tWTR.
3. The RANK system allows more parallelism with lesser timing constraints. It only involves t_{RTR}.

In a FCFS queue we cannot have the option of keeping more than one active command of any requestor. So these commands are propagated every R × Mr times. Due to modified FCFS arbitration, the PRE commands do not get affected by tRRD or tFAW. Here, one more important point is that ACT command of RANK in any moment does not get affected by the PRE command. So if by any reason an ACT command cannot be propagated, then it is safe to send the next PRE command. The above figure shows the implementation of rule 5. Reordering of FCFS queue is not needed if the maximum gap between transitions is tRTR, but when this condition is violated, the entire queue is modified.

BANK partitioning must be available for this design because the back end involves per-request queue. This can be achieved in many ways. One method is to implement it in hardware. Using linker and virtual memory-base partitioning are the other two methods. Data sharing can happen between soft RANKS since they do not have BANK privatization. A different shared BANK partition is used in case of hard requestor sharing a memory with another requestor. Direct memory access is always thought as different requestors, and DMA BANK partition can be accessed when it is not transferring any data.

4 Memory Controller

4.1 RANK-Swapping Mechanism

As opposed to the ideal situation, the data bus utilization is much lower than hundred percent in practical cases. This condition holds true even if all the requestors are open. This is due to the time taken by successive write and read commands which are dependent on t_{RTW} and t_{WTR}. This can be illustrated by the following examples. The initial part of Fig. 4 shows the worst-case condition for different requesters with four open requests in a system where the RANK is one.

To finish all the requests, it will take up to 52 clock cycles. The use of a data bus is limited to 16 cycles, making bus utilization 31%. One can avoid unnecessary timings like t_{RTW} and t_{WTR} if more than one RANK is used. The effect of using different numbers of RANKS can be seen from the above figure. Alternative request by different requestors is assigned to different RANKS. The only timing constraint involved in this scenario is t_{RTR}. This approximately doubles the bus utilization. With RANK 4, bus utilization up to 66% can be achieved. From this, we can conclude that RANK switching alone helps in decreasing the latency without needing the controller's front end.

4.2 Arbitration Rules

Figure 5 shows the memory controllers' back-end design. The main aim is increasing bus utilization and reducing latency bounds of hard requestors. In this situation, every hard or soft requestor gets a RANK but only one. Many requestors can be mapped to single RANK, but no requestor gets two RANKS. The requestors which are assigned the same RANK will get a number of BANKS statistically. For example, let Mr. Requestors use RANK r where $1 \leq r \leq R$. Using BANK privatization, BANKS are statistically partitioned.

Here, RANK is always assumed more than two.

The above diagram describes the command arbitration logic part of the memory controller. In the above figure, different RANKS are assigned to different requestors.

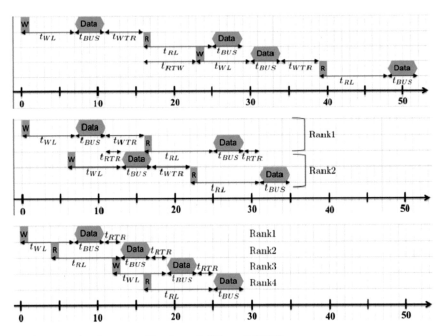

Fig. 4 Comparison between arbitration and different RANKS

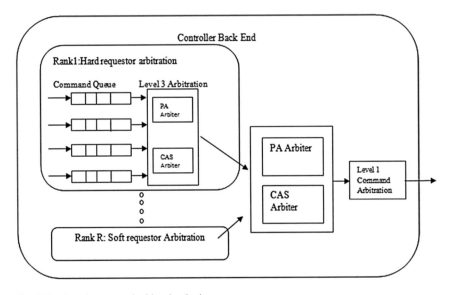

Fig. 5 Back-end command arbitration logic

RANK 1 has been assigned to four hard requestors; therefore, M1 = 4. This logic involves three levels of arbitration, i.e., L1, L2, and L3. The L3 also known as requestor arbitration has separate command queues for hard requestors which are assigned the same RANK. Since RANK 1 is hard to RANK, it can be seen in the figure that it has different command queues. In this level, only requestor's arbitration is done within the same RANK. During arbitration, the L3 arbiter selects the first command of the selected request. Then, that command is sent to L2 which is also known as RANK arbitration; it arbitrates between different RANKS. Arbiters in these levels have two parts PA and CAS. PA handles PRE and ACT commands, and CAS handles read–write commands. L1 arbitration is also known as command arbitration which is the last level. It gives CAS commands higher priority compared to PRE and ACT commands.

Here are some rules which are necessary to satisfy timing constraints of arbitration. These rules are applicable to both level 2 and level 3 arbiters. The behavior will be explained as below.

1. Each requestor has a separate queue. The command which is present at the top of the queue will be called as the active command only if all the previous commands from those requestors are sent, and timing conditions are satisfied. This is true in case of CAS command. This situation becomes active only when the last CAS command's data is transferred. If in a system no other requests are present other than an active command, it is issued immediately.

2. The arbitration policy used by level 3 PA arbiter is modified first come first serve. If a command has active PRE and ACT commands, they are joined immediately to the FIFO queue. Once the level 1 arbiter issues these commands, these are taken out of the queue. This queue is scanned by the arbiter every clock cycle looking for a valid command to be sent first without violating timing constraints. Issuing active PRE command has no problem, but t_{RRD} and t_{FAW} affect issuing the active ACT commands.

3. The level 3 CAS arbiter makes use of standard first come first serve. Only active CAS commands are added to the queue based on the FCFS policy they are issued. Once level 1 arbiter issues them, they are taken out from the queue. For data transmission, the calculation of t_{SDr} is computed by this arbiter and passed to level 2 Arbiter. The t_{SDr} gives an idea of the earliest time for data transfer. The information from the last CAS command is taken to calculate t_{SDr}.

4. In level 2 PA arbiters, any of scheduling policies either FCFS or RR is used.

5. The CAS arbiter at level 2 makes use of modified FCFS. L3 issues command and RANK at the queue of the L2 CAS arbiter. These are removed when this command is issued at L1. The tED gives the idea of when data transfer of previously sent CAS command ends. Every clock cycle tSDr ≤ tED + tRTR is calculated, and the one with the smallest tSDr is propagated.

5 Implementation of RSR

As explained in the above section is the generalized explanation of the RSR design. In the below section, we are going to explain the actual implementation details for the same. The top-level block diagram of the RSR is shown in Fig. 6.

As shown in the below diagram, the external inputs to the memory controller are the clock and the reset. These inputs to the memory controller will generate different outputs which are listed on the right side of the controller. The full memory controller module has been designed using Verilog for FPGA. As shown in Fig. 7 contains the two important sub-modules of the top module shown in Fig. 6.

As shown in Fig. 7, the top-level RSR memory controller contains the two sub-modules as front-end controller and the back-end controller. All the relevant DDR timings can be set using this module. The DDR timing details for a particular ver-

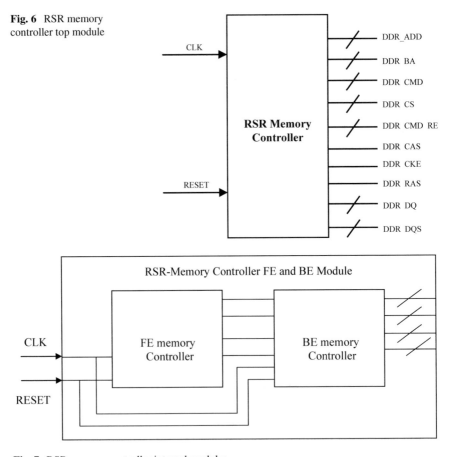

Fig. 6 RSR memory controller top module

Fig. 7 RSR memory controller internal modules

sion of the memory can be taken from the datasheet of that particular device. The front-end memory controller once again will be having one more submodule called as front-end command generator. The front-end part of the memory controller gets the command from the user, whereas the back end of the memory controller interacts with the actual memory device. In the front-end memory controller, we will be having the option of the assignment of different RANKS and BANK addresses w.r.t. different requestors set being selected. The requestors selected can be 16, 8, or any other combinations depending upon our application. The RANKS and the BANKS selected also may vary depending upon the need. Here, different possible combinations can be (RANKS-4, BANKS-4), (RANKS-2, BANKS-8), (RANKS-4, BANKS-2), (RANKS-2, BANKS-2).

We have created a mode register block which contains all the timing parameters for the DDR device. Here, we have included the four main configurations. It also includes the input files used for the requests. Here, we will have the option to select one particular file for a request.

The different possible configurations are listed below:

- REQS = 16, RANKS = 4, BANKS = 4
- REQS = 16, RANKS = 2, BANKS = 8
- REQS = 8, RANKS = 4, BANKS = 2
- REQS = 8, RANKS = 2, BANKS = 4.

6 Simulation Results

Figure 8 shows the simulation result of one of the configurations. We can go on changing the configuration and check different possibilities. The variations in the requests, RANKS, and the BANKS are possible.

7 Evaluation

DDR3 is the device under consideration. For our design, the RSR considered can work with two or four RANKS each containing eight or four BANKS. This gives us the freedom of assigning each request with one BANK. Then, they will be assigned with individual RANK. Four requestors will be assigned for a 16 requestors, 4 RANKS combination. We have used Verilog HDL to implement entire memory controller back end along with the command generator. We have synthesized the design on many of the Xilinx devices, by changing the different combinations of RANKS and BANKS (Fig. 9).

In our design (RSR), the latency depends upon the row hit ratio. The increase in requests and the RANKS increases the performance metrics of this method compared to other methods. The Gem5 is a simulator which can be used to generate various memory traces. We have generated such memory traces and given them as the inputs to the analysis of the memory controller simulation.

8 Synthesis Results

Figure 10 shows the RTL schematic of the front-end and back-end modules of the memory controller. We can see how the front-end and the back-end modules work in tandem to get the work done.

Table 1 presents the synthesis report. Here, we have tried implementing the memory controller with a different combination of RANKS and BANKS. For each of the combination, we have also used a different set of reconfigurable devices as the platforms to check its performance.

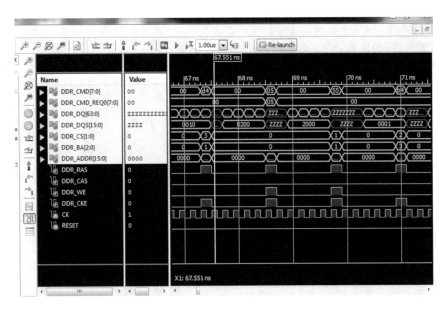

Fig. 8 Simulation results of the memory controller module

Fig. 9 8 Requestors each having 64 bits of data bus

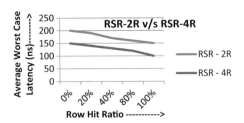

Table 1 Synthesis report generated once we run the code using different devices and different combinations of requests, RANKS, and BANKS

Devices	Number of slice registers	Number of slice LUTs	Number used as memory	Number of unique control sets	Timing report (ns)
	Requests_8-RANKS_4-BANKS_2				
6vcx75tff484-2 (Virtex-6)	3106 out of 93,120 (3%)	278,014 out of 46,560 (597%)	212,992 out of 16,720 (1273%)	326	3.597
	Requests_8-RANKS_2-BANKS_4				
6vcx75tff484-2	2463 out of 93,120 (2%)	276,779 out of 46,560 (594%)	212,992 out of 16,720 (1273%)	271	3.589
	Requests_16_RANKS_4_BANKS_4				
6vcx75tff484-2	4609 out of 93,120 (4%)	553,704 out of 46,560 (1189%)	425,984 out of 16,720 (2547%)	518	3.590
	REQS_16_RANKS_2_BANKS_8				
6vcx75tff484-2	4360 out of 93,120 (4%)	553,391 out of 46,560 (1188%)	425,984 out of 16,720 (2547%)	479	3.590

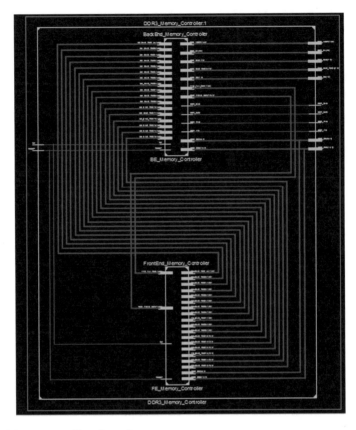

Fig. 10 Memory controller schematic

9 Conclusion

A RANK-swapping unblocked row DRAM controller is a newly designed method for DDR DRAM. The currently available memory controllers depend on static command schedules. This design considers the dynamic nature of DRAM. In this design, bus utilization is guaranteed to increase by using data bus within RANK-to-RANK switch time. It is much lesser compared to read to write delay or vice versa. One more advantage of this design is to provide isolation between different requestors.

Acknowledgements I would like to thank Dr. Hansraj Guhilot for his motivation and valuable suggestions. I would also thank the principal, HOD E and C, and all the teaching and non-teaching staff of KLE Dr. M.S.S. CET, Belagavi, for their support in completing this work.

References

1. Bui D et al (2011) Temporal isolation on multiprocessing architectures. In: 48th ACM/EDAC/IEEE design automation conference (DAC), 2011. IEEE
2. Wu Z (2013) Worst case analysis of DRAM latency in hard real-time systems. MASc thesis. University of Waterloo
3. Paolieri M, Quinones E, Cazorla F, Valero M (2009) An analyzable memory controller for hard real-time CMPs. IEEE Embed Syst Lett 1(4):86–90
4. Akesson B, Goossens K, Ringhofer M (2007) Predator: a predictable SDRAM memory controller. In: 5th IEEE/ACM/IFIP international conference on hardware/software codesign and system synthesis (CODES+ ISSS), 2007. IEEE
5. Ecco L, Tobuschat S, Saidi S, Ernst R (2014) A mixed critical memory controller using bank privatization and fixed priority scheduling. In: Proceedings of the 20th IEEE international conference on real-time computing systems and applications (RTCSA), Aug 2014
6. Wang DT (2005) Modern DRAM memory systems: performance analysis and scheduling algorithm. Ph.D. dissertation, The University of Maryland at College Park
7. Kim S, Soontae K, Lee Y (2012) DRAM power-aware rank scheduling. In: Proceedings of the 2012 ACM/IEEE international symposium on low power electronics and design. ACM
8. JEDEC (2012) DDR3 SDRAM Standard JESD79-3F, July 2012
9. Reineke J et al (2011) PRET DRAM controller: bank privatization for predictability and temporal isolation. In: Proceedings of the 9th IEEE/ACM/IFIP international conference on hardware/software codesign and system synthesis (CODES+ISSS), 2011, pp 99–108
10. Bourgade R et al (2008) Accurate analysis of memory latencies for WCET estimation. In: 16th international conference on real-time and network systems (RTNS 2008)

Image Enhancement Based on Fractional Calculus and Genetic Algorithm

G. Sridevi and S. Srinivas Kumar

Abstract Image enhancement is an interesting topic in the image processing area. In this work, image enhancement with fractional-order derivative and genetic algorithm is proposed. Fractional-order derivative possesses a non-local property, which is helpful to find the fine edges of the image. In this paper, firstly, fractional-order partial differences are computed in forward x-direction, backward x-direction, forward y-direction, and backward y-direction. These differences are represented based on discrete Fourier transform (DFT). Finally, genetic algorithm (GA) is applied for the fractional-order selection to get optimum results and the fractional-order is chosen in the range from 0 to 1. The experimental results give the superiority of the proposed algorithm than the traditional methods.

Keywords Fractional-order derivative · Fourier transform · Genetic algorithm · Image enhancement

1 Introduction

Image enhancement is a technique to get better quality of the image when the captured image undergoes bad visual effect due to the natural condition from the environment and the limitations of the equipment. This technology has gradually been involved in every aspect of human life. The related social production has been widely used in aerospace, biomedical, industrial production and public security and other fields. Potential applications include highlighting the characteristics of target in the image, extracting the target feature from the digital image, and so on, which are conducive to target identification, tracking, and image understanding.

G. Sridevi (✉)
Department of ECE, Aditya Engineering College, Surampalem, AP, India
e-mail: sridevi_gamini@yahoo.com

S. Srinivas Kumar
Department of ECE, JNTUK, Kakinada, AP, India

© Springer Nature Singapore Pte Ltd. 2019
N. Chaki et al. (eds.), *Proceedings of International Conference on Computational Intelligence and Data Engineering*, Lecture Notes on Data Engineering and Communications Technologies 28, https://doi.org/10.1007/978-981-13-6459-4_20

The image enhancement technology can be arranged into two groups based on different processing spaces frequency-domain approaches and spatial domain approaches. The first one regards the image as a 2D signal and employs the 2D discrete Fourier transform (DFT) to enhance the signal [1]. By using the low-pass filtering methods, the noise in the images can be removed and the high-pass filtering methods can enhance the high-frequency signals such as edges and textures and make the blurry image becomes clear. The frequency-domain methods are indirect enhancement methods, while the spatial domain methods are direct.

The spatial domain methods are again classified into the point operation methods and the neighborhood operation methods. The point operation methods include the grayscale correction, gray transform, and the histogram modification. Its purpose is to uniform the image or expand the dynamic range of the image. There are two kinds of neighborhood-based enhancement algorithms, i.e., image smoothing and sharpening. Smoothing algorithms are generally used to remove the image noise, but it also caused the edge to be fuzzy. The commonly used algorithms are the average filtering and median filtering. Sharpening algorithms are used to highlight the edge contour of the object for target identification.

The commonly used algorithms are the gradient methods, high-pass filtering, mask matching methods, and statistical difference methods, etc. They all can be used to remove or reduce noise [2, 3]. The biggest advantage of the average filtering and median filtering is simple and the speed is faster, the disadvantage is that it will cause texture details and edge to blur. In a word, after dealing with the neighborhood-based algorithm, the noise intensity of the image is weak. However, along with the increase of the neighborhood, the image will become more and fuzzier and lose some edge details. The low-pass filtering algorithm can keep the low-frequency information better, but suppresses the high-frequency information of the local details too much.

In recent years, scholars had made immense development in the fractional calculus theory and have been extensively practiced in many fields such as the chemical, electromagnetic science and control science, materials science and so on. Recently, the fractional-order calculus theory has become a hot research issue and has been applied to image enhancement in recent years. Compared to the integral order calculus, the implementation of the fractional-order calculus is more complex, but it extends the order of the differential operator and has more freedom degree and flexibility. Furthermore, the fractional differential can enhance the texture, edges, and preserve the flat regions of the image. When processing these features, the traditional fractional differentials can construct the masks with the same value in eight directions. However, under such circumstances, the edge, weak textures, and flat regions would be unnoticed. One promising way is to use the fractional differential for enhancing image to obtain more clear texture with high vision quality.

Mathieu et al. [4] proposed the fractional-order differential edge detection operator, namely fractional robust profile edge detector (CRONE). When the fractional differential order is in the range of [1, 2], it can selectively detect the edges of the image also this detection operator can overcome the noise effect in the process of edge detection. Panetta et al. [5] proposed the fractional differential filter and applied it to edge detection. By using the basic theory of fractional calculus, they designed a differential order filter which can adjust the order of the differential and thus solve the drift problem of the traditional edge detection operator and suppress some noise.

By studying the basic definition of fractional calculus, and the realization method of the corresponding fractional-order differential operator, Li and Xie [6] introduced the gradient features of the image and the human visual characteristics to the fractional differential operator so as to construct an adaptive fractional-order differential image enhancement model. Sanjay et al. [7] explain a new class of edge detection algorithm derived from the fractional-order differential operator and through the usage of the fractional Fourier transform.

Recently, Pu et al. [8, 9] utilized the fractional differential filters for the enhancement of the images. The authors constructed a mask based on fractional differential and it was proved that it had better enhancement effect than the integral order calculus. Yu et al. [10, 11] proposed the fractional differentiation-based method for the enhancement of textures in image processing. The authors developed the masks in eight directions based on Riesz definition. Sridevi and Kumar [12] utilized fractional-order derivative based in Caputo definition for image inpainting and enhancement.

Saadia and Rashdi [13] introduced an adaptive image enhancement technique that chooses the fractional-order for every pixel automatically. This method selected higher fractional-order when the magnitude of the gradient is large and smaller order for little magnitude of gradient. Gao et al. [14] generalized the fractional differential operators by applying these operators to quaternion and developed a set of masks which was called the quaternion fractional differential (QFD) operators.

However, there exist some unsolved problems in image enhancement. The traditional fractional differential operators construct the masks which have the same value in the eight directions and ignore the surrounding information (such as the image edge, clarity, and texture information) and texture of the pixels. Consequently, it is significant to present a novel image enhancement algorithm which considers the surrounding information (such as the image edge, clarity, and texture information) and preserves the texture simultaneously.

In order to deal with this problem and enhance the image effectively, in the development of image enhancement, why the image quality degrades is not addressed by us. This article explains a novel algorithm on image enhancement which depends on the fractional differential and the genetic algorithm. This method makes use of the surrounding information (such as the image edge, clarity, and texture information) and structural features of different pixels, and considers the genetic algorithm for the optimal fractional-order. The competitive image enhancement results were produced in comparison with the conventional methods in the proposed algorithm.

This article is outlined as follows. In Sect. 2, the discrete Fourier transform (DFT) based fractional-order derivative is discussed. Genetic algorithm is explained in Sect. 3. In Sect. 4, the proposed method for image enhancement is discussed. Experimental results are illustrated in Sect. 5 and conclusions are presented in Sect. 6.

2   Fractional-Order Derivative

The fractional-order (non-integer) derivative is the generalized form of the integer-order derivative. It is calculated in various approaches and the well-known definitions that have been familiarized are the Caputo [12], Riemann–Liouville [8] and Grunwald–Letnikov [8] definitions. On the other hand, in this article, the frequency-domain definition [2] is used as it is simple to implement. For any function $f(t) \in L_2(R)$ the Fourier transform of it is

$$\hat{f}(\omega) = \int_{-\infty}^{\infty} f(t) \exp(-j\omega t)dt \tag{1}$$

The equivalent form of the first-order derivative in the frequency domain is

$$Df(t) \leftrightarrow (j\omega)\hat{f}(\omega) \tag{2}$$

where "$\leftrightarrow$" denotes the Fourier transform pair. The 2D DFT of an image $u(x, y)$ of $M \times M$ size is

$$\hat{u}(\omega_1, \omega_2) = \frac{1}{M^2} \sum_{x,y=0}^{M-1} u(x, y)e^{-\frac{j2\pi(\omega_1 x + \omega_2 y)}{M}} \tag{3}$$

For 2D DFT, the shifting property in spatial domain can be denoted as

$$u(x - x_0, y - y_0) \overset{F}{\leftrightarrow} e^{-\frac{j2\pi(\omega_1 x_0 + \omega_2 y_0)}{M}} \hat{u}(\omega_1, \omega_2) \tag{4}$$

Thus, the first-order partial difference in x-direction can be denoted as

$$D_x u(x, y) = u(x, y) - u(x - 1, y) \tag{5}$$

$$D_x u(x, y) \overset{F}{\leftrightarrow} \left(1 - e^{-\frac{j2\pi\omega_1}{M}}\right)\hat{u}(\omega_1, \omega_2) \tag{6}$$

and the DFT of fractional-order (α) partial difference in x-direction can be denoted as

$$D_x^a u(x, y) \overset{F}{\leftrightarrow} (1 - e^{-\frac{j2\pi\omega_1}{M}})^a \hat{u}(\omega_1, \omega_2) \tag{7}$$

where, F is 2D-DFT. Similarly, the DFT of fractional-order partial difference in y-direction can be denoted as

$$D_y^a u(x, y) \overset{F}{\leftrightarrow} (1 - e^{-\frac{j2\pi\omega_2}{M}})^a \hat{u}(\omega_1, \omega_2) \tag{8}$$

In general, the fractional-order derivative operator of 2D signal can be denoted as

$$D^a u(x, y) = \left(D_x^a u(x, y), D_y^a u(x, y) \right) \tag{9}$$

and

$$\left| D^a u(x, y) \right| = \sqrt{(D_x^a u(x, y))^2 + (D_y^a u(x, y))^2} \tag{10}$$

In practical computations, to calculate the fractional-order difference, the central difference method is very useful. This is equivalent to shifting $D_x^a u(x, y)$ and $D_y^a u(x, y)$ by $a/2$ units.

$$\tilde{D}_x^a u(x, y) = F^{-1}\left(\left(1 - e^{-\frac{j 2\pi \omega_1}{M}} \right)^a e^{\frac{j\pi a \omega_1}{M}} \hat{u}(\omega_1, \omega_2) \right) \tag{11}$$

$$\tilde{D}_y^a u(x, y) = F^{-1}\left(\left(1 - e^{-\frac{j 2\pi \omega_2}{M}} \right)^a e^{\frac{j\pi a \omega_2}{M}} \hat{u}(\omega_1, \omega_2) \right) \tag{12}$$

where F^{-1} is the 2D inverse discrete Fourier transform (IDFT). The operator $\tilde{D}_x^a$ has the form $\left[F^{-1} \right][K_1][F]$, where [.] is a matrix operator, and

$$K_1 = \text{diag}\left(\left(1 - e^{-\frac{j 2\pi \omega_1}{M}} \right)^a e^{\frac{j\pi a \omega_1}{M}} \right) \tag{13}$$

The adjoint operator $\tilde{D}_x^{a*}$ of $\tilde{D}_x^a$ can be computed by the following conception

$$\tilde{D}_x^{a*} = \left(\left[F^{-1} \right][K_1][F] \right)^* = \left[F^{-1} \right]^* \left[K_1^* \right][F]^* = [F]\left[K_1^* \right]\left[F^{-1} \right] \tag{14}$$

Since K_1 is purely diagonal operator, K_1^* is the complex conjugation of K_1. The same procedure can be used for the calculations of $\tilde{D}_y^{a*}$ and $\tilde{D}_y^a$.

$$\tilde{D}_y^{a*} = [F]\left[K_2^* \right]\left[F^{-1} \right] \tag{15}$$

3 Genetic Algorithm

Genetic algorithm (GA) [15] is a heuristic approach to achieve constructive solutions by experimental analysis. In order to optimize the problem, GA searches for a solution by the methods mimicking natural biological progression, for example, hereditary, selection, crossover, and mutation. The GA is initiated by generating random population. Next, GA evaluates each candidate in the population in accordance with the fitness function. Lastly, GA strives to minimize or maximize the result obtained by the fitness function.

A population of random variables is considered as initial population. The corresponding fitness function is evaluated for each variable and it is repeated for M

number of generations. Next, from these generations, the strongest fitness individuals are stochastically selected and become parents for the next generation. In order to change the features of subsequent generations, GA applies crossover and mutation operators on the selected parents. Finally, this process ends when the algorithm produces a satisfactory fitness level has been reached for the population or maximum number of generations.

The major components used in GA are

a. **Selection**: It is related to Darwinian principle of survival of the fittest. Accordingly, for the next generation only the strong population survives and weak ones are vanished. Likewise, in the process of selection, low fitness individuals are eliminated from the population of size N, N_{good} individuals survive for mating and the leftover $N - N_{good}$ are discarded to make room for offspring in the next generation.

b. **Crossover**: It provides two new offsprings to select two parents from N_{good} individuals. A crossover point is selected between the first and last element of chromosomes of parents. In order to produce new two offspring, the tails of parents from the crossover point are interchanged. Based on the crossover probability, the parents' population will participate in the crossover operation.

c. **Mutation**: It produces new features in the offspring which are different from their priors. Hence, mutation recommends genetic diversity in population.

4   Proposed Model

The gradient of the image in the smooth or nearly flat regions is zero, whereas the fractional gradient produces nonzero value. So the fractional-order derivative operator is very helpful to enhance the smooth and texture regions of the image. In this paper, the fractional-order derivative is computed based on discrete Fourier transform since it is very simple to implement [2] (Fig. 1).

The fractional-order is not the same to enhance the different images. In traditional image enhancement methods, the fractional-order selection is done by manually. In this work, the fractional-order is selected by using genetic algorithm. The fitness function for this model is based on information entropy (IE) [16] of the enhanced image which is given by the Eq. (16)

$$H(u) = - \sum_{i=0}^{L_{max}-1} p_u(i) \log_2(p_u(i)) \tag{16}$$

where $p_u(i)$ is the probability of occurrence of pixels with ith intensity value in the enhanced image u and L_{max} is the maximum intensity value present in the enhanced image.

The following steps are implemented for the enhanced image

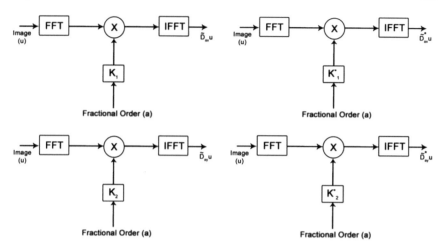

Fig. 1 Calculation of fractional-order partial differences

1. Compute DFT of the given image u, i.e., $\hat{u}$
2. Compute K_1, K_1^*, K_2, K_2^*
3. Compute matrix multiplications of $\hat{u}$ and K_1, $\hat{u}$ and K_1^*, $\hat{u}$ and K_2, $\hat{u}$ and K_2^*
4. Compute the IDFT of results of step 3, i.e., $\tilde{D}_x^a u$, $\tilde{D}_x^{a^*} u$, $\tilde{D}_y^a u$, $\tilde{D}_y^{a^*} u$, respectively
5. Compute edges of the image, $u_E = \tilde{D}_x^a u + \tilde{D}_x^{a^*} u + \tilde{D}_y^a u + \tilde{D}_y^{a^*} u$
6. Compute enhanced image by using $u_E + u$
7. Compute the Information entropy of the enhanced image
8. Compute fitness function $=$ Information entropy
9. Apply genetic algorithm to choose fractional-order for optimizing the fitness function.

5 Experimental Results and Discussion

In this section, to evaluate the proposed technique against the traditional fractional differential algorithm, a set of experiments are conducted. Also, to explain the performance of the proposed method some benchmark images have been used. The images are taken from USC-SIPI image database [17]. The presented method was implemented by MATLAB on PC computer with 1.6 GHZ CPU and 4 GB RAM.

In traditional approaches, the selection of the fractional-order is done by manually and this fractional-order is varied in accordance with the characteristics of the image. This fractional-order selection process is done automatically in the proposed method by means of GA. The number of generations is selected in this algorithm are 10 and 10 number of random population is selected in the range [0, 2]. Here, crossover probability of 0.8 is selected and mutation probability of 0.1 has been done. The fitness function is the information entropy (IE) of the image.

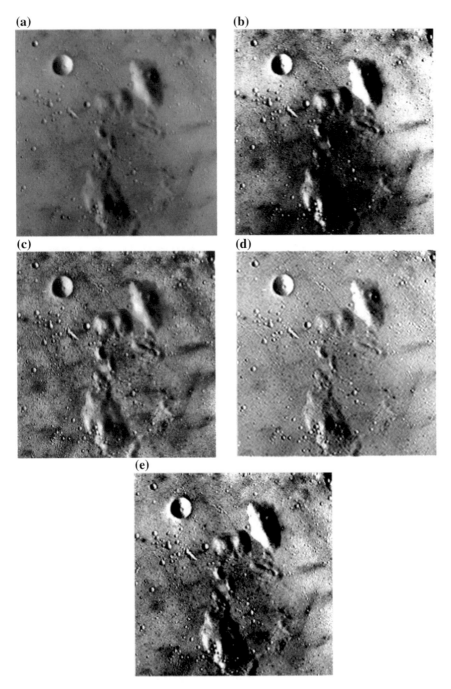

Fig. 2 a Original image. **b** Result of histogram equalization. **c** Result of method [8]. **d** Result of method [8] with GA. **e** Proposed model

Table 1 Comparison of image enhancement techniques on various images

Image	IE of input image	Method [8]		Method [8] with GA		Proposed model	
		a	IE	a	IE	a	IE
Moon's surface	6.7093	0.5	7.1511	0.4389	7.2214	**0.1623**	**7.5262**
Bridge	5.7056	0.5	7.6929	0.5249	7.7455	**0.8866**	**7.8176**
Lena	7.4456	0.5	7.7562	0.4868	7.7721	**0.6911**	**7.8049**
Mandrill	7.3583	0.5	7.6962	0.4995	7.6963	**0.8912**	**7.6963**
Barbara	7.4664	0.5	7.7821	0.4819	7.7960	**0.7488**	**7.7930**
Peppers	7.5937	0.5	7.8058	0.4888	7.8235	**0.8368**	**7.6987**

The experimental results are displayed in Fig. 2, which indicates the proposed algorithm not only enhances the image quality, but also preserves smooth area and weak texture contains together global and local information in the image. The best order of the fractional differential is $a = 0.1623$, i.e., information entropy is maximum. The simulation results on various images have been shown in Table 1, which are indicated in Bold face. The proposed model can able to enhance the texture as well as smooth regions effectively by selecting the fractional-order automatically.

6 Conclusions

In this work, the fractional-order derivative in the directions of positive and negative x- and y-axes based on discrete Fourier transform. These four results are added to get the edge information. This result is added to the original image for the enhanced result. The fractional-order is selected by using genetic algorithm to optimize the fitness function. The proposed model is also compared with the GL-based enhancement method.

References

1. Chan R, Lanza H (2013) An adaptive strategy for the restoration of textured images using fractional order regularization. Numer Math Theory Methods Appl 6:276–296
2. Jian B, Xiang CF (2007) Fractional-order anisotropic diffusion for image denoising. IEEE Trans Image Process 16:2492–2502
3. Pu YF, Yuan X, Liao K, Zhou J (2005) Structuring analog fractance circuit for 1/2 order fractional calculus. In: Proceedings of the IEEE 6th international conference, ASIC, vol 2, pp 1039–1042
4. Mathieu B, Melchior P, Oustaloup A, Ceyral C (2003) Fractional differentiation for edge detection. Sig Process 83:2421–2432

5. Panetta KA, Wharton EJ, Agaian SS (2008) Human visual system-based image enhancement and logarithmic contrast measure. IEEE Trans Syst Man Cybern B Cybern 38:174–188
6. Li B, Xie W (2015) Adaptive fractional differential approach and its application to medical image enhancement. Comput Electr Eng 45:324–335
7. Kumar S, Saxena R, Singh K (2017) Fractional Fourier transform and fractional-order calculus-based image edge detection. Circ Syst Signal Process 36:1493–1513
8. Pu YF, Liu Z, Xiao Y (2010) Fractional differential mask: a fractional differential-based approach for multi-scale texture enhancement. IEEE Trans Image Process 19:491–511
9. Pu YF, Wang W, Zhou J, Wang Y, Jia HD (2008) Fractional differential approach to detecting textural features of digital image and its fractional differential filter implementation. Sci China Ser 51:1319–1339
10. Yu Q, Liu F, Turner I, Burrage K, Vegh V (2012) The use of a Riesz fractional differential-based approach for texture enhancement in image processing. ANZIAM J 54:590–607
11. Yu Q, Vegh V, Liu F, Turner I (2015) A variable order fractional differential-based texture enhancement algorithm with application in medical imaging. PLoS One 10(7):e0132952
12. Sridevi G, Kumar SS (2017) Image inpainting and enhancement using fractional order variational model. Def Sci J 67:308–315
13. Saadia A, Rashdi A (2016) Echocardiography image enhancement using adaptive fractional order derivatives. In: IEEE international conference on signal and image processing, pp 166–169
14. Gao CB, Zhou JL, Hu JR, Lang FN (2011) Edge detection of colour image based on quaternion fractional differential. IET Image Proc 5:261–272
15. Meenakshi K, Rao CS, Prasad KS (2014) A hybridized robust watermarking scheme based on fast Walsh-Hadamard transform and singular value decomposition using genetic algorithm. Int J Comput Appl 108(11):1–8
16. Sarangi P, Mishra B, Majhi B, Dehuri S (2014) Gray-level image enhancement using differential evolution optimization algorithm. In: International conference on signal processing and integrated networks, pp 95–100
17. http://sipi.usc.edu/database

Cepstrum-Based Road Surface Recognition Using Long-Range Automotive Radar

Sudeepini Darapu, S. M. Renuka Devi and Srinivasarao Katuri

Abstract During driving, a sudden change in the road surface results in imbalance of vehicle due to wheel slip which leads to accidents. Thus, a need arises for an automotive system to recognize the type of road surface ahead and alert the driver to accordingly change the speed of the vehicle. This paper proposes a technique for road surface recognition using 77 GHz frequency-modulating continuous wave (FMCW) long-range automotive radar. The cepstral coefficients calculated from the backscattered signal are analyzed, using classifiers like decision tree and SVM. This technique recognizes five different road surfaces, i.e., dry concrete, dry asphalt, slush, sand, and bushes. To validate the accuracy and classification rate, field testing is conducted at Kondapur (Telangana) and the system has achieved prediction percentage of above 90%.

Keywords Automotive systems · Radar backscattering · Signal processing · Cepstral coefficients · Machine learning · Road surface detection

1 Introduction

Over the past few years, advanced driver-assistance Systems (ADAS) have been a significant area of research. The reason being, ADAS includes many novel applications such as detecting guardrail and vehicle [1], collision avoidance [2], detecting pavement and lane boundaries [3], and automatic cruise control [4] that are to be the part of modern cars. Nowadays radars are used in the above applications as a

S. Darapu (✉) · S. M. Renuka Devi
G Narayanamma Institute of Technology and Science, Shaikpet, Hyderabad, India
e-mail: reddy.sudeepini@gmail.com

S. M. Renuka Devi
e-mail: renuka.devi.sm@gmail.com

S. Katuri
Ineda Systems Pvt. Ltd., Kothaguda, India
e-mail: srinivasarao.katuri@inedasystems.com

© Springer Nature Singapore Pte Ltd. 2019
N. Chaki et al. (eds.), *Proceedings of International Conference on Computational Intelligence and Data Engineering*, Lecture Notes on Data Engineering and Communications Technologies 28, https://doi.org/10.1007/978-981-13-6459-4_21

standard equipment to facilitate safety and comfort driving. These automotive radars are allocated with two frequency bands, i.e., 22–24 GHz (short range up to 30 m), 76–77 GHz (long range up to 150 m). Presently, road surface type can be recognized using some mechanical devices and optical sensors [5]. But this has limited detection range and too expensive for medium and lower class cars. But multipurpose radar sensors provide substantial detection range and yields cost benefits. Bystrov et al. proposed a technique using radar in which backscattered signals are analyzed, dissimilarity in the features from different surfaces are found, and finally, clustering algorithm is used for classification of road surfaces [6]. Hakli et al. developed measurement system for examining the backscattering properties and to estimate the surface friction of road using 24 GHz automotive radar [7].

Viikari used a 24 GHz automotive radar technology to recognize road conditions [8] by measuring the low-friction spots of the backscattered signal at different polarizations to recognize ice, water and snow, and asphalt.

Raj et al. have developed a system based on image processing by analyzing the texture properties to detect six road surfaces such as cement road, bushes road, sandy road, asphalt road, rough road, and rough asphalt road [9].

Recognizing different road conditions using bistatic radar operating at 61 GHz is presented by Kees and Detlefsen [10]. Under the car chassis, the radar system is located which can transmit and receive at both linear polarizations. All the four polarization combinations are coherently measured by the system and by relating the phases and amplitudes between different polarizations, road surfaces such as cobblestone and asphalt pavements are recognized. Patents on road surface recognition using automotive radars are presented in [11].

In this paper, capability of 77 GHz FMCW automotive radar in recognizing road surfaces using cepstrum analysis on backscattered signals is studied. Novelty of the proposed system is using cepstral analysis for backscattered signals from radar. Cepstrum is a technique which is used for speech signals for classification because speech signals are FM (Frequency Modulated) signals. We have considered cepstrum processing in the proposed method because we are using FMCW radar which generates FM signals. Advantages of using 77 GHz FMCW Radar technology is, the detection range is up to 200 m, yields cost benefits and resistant to adverse weather conditions. The paper is organized as follows: Sect. 2 briefs on measurement setup for data collection or field testing and process of data collection. Section 3 describes about process flow of the proposed system and novel approach of using cepstrum for road surface recognition with supervised classification. MATLAB simulation results and field testing prediction accuracies of different road surfaces in real conditions are presented in Sect. 4 and finally conclusion is discussed in Sect. 5.

2 Measurement Setup

Radar is mounted in front of the car and it is configured to cover the range of 20m. Surface recognition is practically possible if the power reflected from the road surface is sufficient to recognize the road type. Energy of the backscattered signals also depends on surface irregularities, surface roughness, and texture of the surface. This measurement setup is used for data collection and field testing in real conditions (Fig. 1).

2.1 Data Collection

Figure 2 shows the block diagram of FMCW radar. Although single chain transmitter and receiver of the radar are shown in figure, the actual system uses single transmitter and four receiver antennas for analyzing backscattered signals.

Local oscillator generates the chirp which is a ramp signal of linear frequency, and modulator modulates the ramp signal and is transmitted by the transmitter antenna (TX). Receiver antenna (RX) receives the signal and mixer is used to generate intermediate frequency (IF) signal by mixing transmitted signal with received signal. The generated IF signal is passed through analog-to-digital Converter (ADC) for digitizing and the digitized signal is used for digital signal processing.

Fig. 1 Measurement set for surface recognition

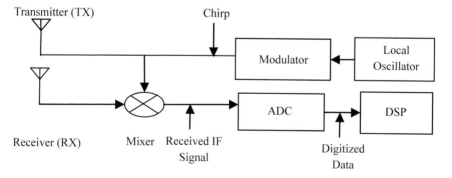

Fig. 2 Block diagram of FMCW radar

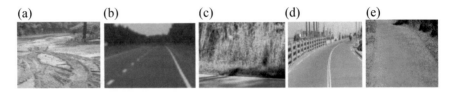

Fig. 3 Image of road surfaces. **a** Slush. **b** Dry asphalt. **c** Bushes. **d** Dry concrete. **e** Road. **e** Sandy surface

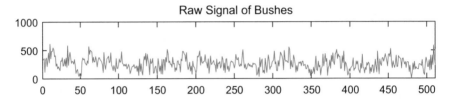

Fig. 4 Raw signal of Bushes

For five different road surfaces such as slush, dry asphalt, bushes, dry concrete and sandy as shown in Fig. 3, data is collected with radar installed by moving the vehicle with speed of around 20 Kmph and covering a total distance of about 10 km. Two sets of data are collected, one for training and the other for testing. For training, huge amount of data is collected from each surface at different places in order to avoid overfitting. Each frame of data is in the form of $128 \times 512 \times 4 \times 2$, where 128 represents number of chirps, 512 represents number of samples in each chirp, 4 represents number of receivers and 2 denotes In-phase (I) and Quadrature-phase (Q) components. Sample plot of raw ADC data for bushes is shown in Fig. 4.

3 Methodology of Proposed System

Block-level implementation of the proposed system is shown in Fig. 5. In training phase, cepstrum processing is done on backscattered signals for the purpose of feature extraction. Reduced set of features which are well differentiating road surfaces are selected and trained using classifiers and the generated model is stored. Processing steps in training and testing are same up to feature extraction. In testing phase after feature extraction, they are fed to trained model. Trained model and extracted features from testing are used to predict to the type of road surface.

Training Phase

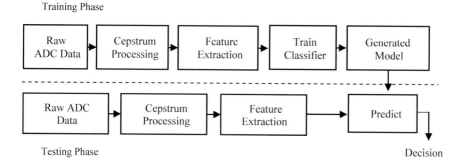

Fig. 5 Training and testing phase for road surface recognition

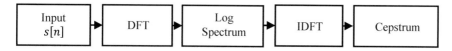

Fig. 6 Process flow of cepstrum processing

3.1 Cepstral Processing and Feature Extraction

Cepstrum is a powerful technique used in many applications like speech processing, radar signal processing, image processing, etc. [12]. Process flow of cepstrum processing is shown in Fig. 6.

$$\text{Power Cepstrum} = \left| F^{-1}\left\{ \log\left(|F\{s[n]\}|^2 \right) \right\} \right|^2 \tag{1}$$

where $s[n]$ is the input signal, F is the fast Fourier transform (FFT) and F^{-1} is the inverse FFT. Magnitude of the processed signal at every sample is called cepstral coefficients. We get 512 cepstral coefficients, as the input data set is of 512 samples. These cepstral coefficients are used as the features. Figure 7, shows FFT of signal, log of the spectrum and cepstrum of signal for one frame of data which is collected from bushes.

3.2 Feature Selection and Classification

Among all cepstral coefficients, only few are considered as features which can well differentiate different surfaces when used for classification. Figure 8 is a graph plotted between number of cepstral coefficients in x-axis and prediction percentage in y-axis using SVM classifier. From figure, it is evident that considering 1–20 cepstral coefficients as features gives best results, beyond this prediction percentage decreases. Classifiers used for classification are decision tree and SVM.

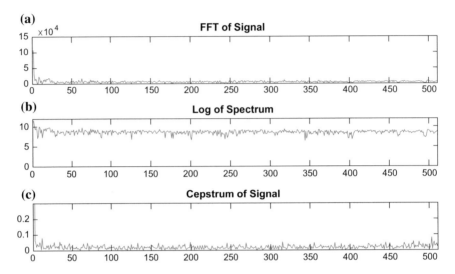

Fig. 7 Plots of cepstral processing for bushes. **a** FFT of signal. **b** Log of spectrum. **c** Cepstrum

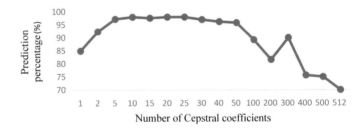

Fig. 8 Number of features and prediction percentage

4   Experimental Results

In this technique, cepstral processing, feature selection, training, and testing is performed using MATLAB. Data set is processed and classifier is trained to generate model for classifiers such as decision tree and SVM. As seen in Sect. 3.2, selected features are 1–20 cepstral coefficients. These 20 features are analyzed and it is observed that cepstral coefficients 1 and 3 are well separating the road surfaces. Scatterplot is drawn by considering cepstral coefficient 1 on x-axis and cepstral coefficient 3 on y-axis for the training data set using SVM classifier, and prediction percentage of about 96.7% is obtained. Using the training data set with 5-folded, a classifier model is generated. The confusion matrix obtained for each of the classifiers, i.e., decision tree and SVM is given in Tables 1 and 2, respectively. From tables, it is observed that SVM gives the best prediction percentage compared to decision tree. Prediction percentage is calculated using (2) (Fig. 9).

Table 1 Confusion matrix using decision tree

Types of roads	Slush surface (%)	Dry asphalt (%)	Bushes (%)	Dry concrete (%)	Sand road (%)
Slush surface	98		1		1
Dry asphalt	1	88	2		9
Bushes	1	2	97		
Dry concrete				99	1
Sandy road	2	7		3	88

Table 2 Confusion matrix using SVM

Types of roads	Slush surface (%)	Dry asphalt (%)	Bushes (%)	Dry concrete (%)	Sand road (%)
Slush surface	94	2	1		3
Dry asphalt		95		1	4
Bushes		2	98		
Dry concrete				100	
Sandy road		2			98

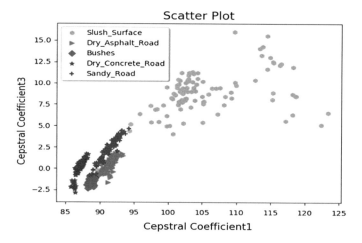

Fig. 9 Scatter plot using cepstral coefficient 1, 3 for SVM classifier

Table 3 Prediction percentage with different classifiers

Types of roads	Slush surface (%)	Dry asphalt road (%)	Bushes (%)	Dry concrete surface (%)	Sand road (%)
Decision tree	96	90	93	90	92
SVM	99	93	98	92	95

$$\text{Prediction Percentage} = \frac{\text{Number of Correctly Predicted Samples}}{\text{Total Number of Training Samples}} \times 100 \qquad (2)$$

Having the model generated, testing data set is used for classification by processing the steps involved in Fig. 5. Prediction percentage is a measure of classifier performance. Table 3 gives information about prediction percentage of testing data with generated trained model. From table, it is observed that SVM performs well with minimum prediction percentage of above 90%.

MATLAB code is converted to python code and the training model is generated using the classifiers. With the generated model, field testing is performed at Kondapur (Telangana) area. The performance of the proposed system in terms of classification accuracy is around a minimum of 90 and 92% for decision tree and SVM, respectively as shown in Table 3.

5 Conclusion

Automatic road surface detection plays an important role in reducing accidents, by warning the driver about the type of the road ahead. In our proposed system, backscattered signal is obtained from automotive FMCW radar of 77 GHz installed in front of the vehicle, then the signal is processed to extract cepstral coefficients and train the classifier model. During testing, prediction accuracy of above 90% is obtained using SVM classifier. Field testing is performed in the area of Kondapur (Telangana) and classification accuracy around 90% is obtained. In the future, we will explore with additional features, different surfaces, and different classifiers, to obtain better prediction accuracy.

Acknowledgements I sincerely thank INEDA SYSTEMS Pvt. Ltd (www.inedasystems.com) and express my gratitude to the officials for their guidance and encouragement in carrying out this project.

References

1. Alessandretti G, Broggi A, Cerri P (2007) Vehicle and guard detection using radar and vision data fusion. IEEE Trans Intell Transp Syst 8(1):95–105
2. Eidehall A, Pohl J, Gustafsson F, Ekmark J (2007) Toward autonomous collision avoidance by steering. IEEE Trans Intell Transp Syst 8(1):84–94
3. Ma B, Lakshmanan S, Hero AO III (2000) Simultaneous detection of lane and pavement boundaries using model-based multisensor fusion. IEEE Trans Intell Transp Syst 1(3):135–147
4. Abou-Jaoude R (2003) ACC radar sensor technology, test requirements, andtestsolutions. IEEE Trans Intell Transp Syst 4(3):115–122
5. Andersson M, Bruzelius F, Casselgren J, Gafvert M, Hjort M, Hultén J, Habring F, Klomp M, Olsson G, Sjodahl M, Svendenius J, Woxneryd S, Walivaara B (2007) Road friction estimation. IVSS project report. Saab Automobile AB, Trollhattan, Sweden
6. Bystrov A, Abbas M, Hoare E, Tran T-Y, Clarke N, Gashinova M, Cherniakov M (2014) Remote road surface identification using radar and ultrasonic sensors. In: Proceedings IEEE European radar conference, pp 185–188
7. Hakli J, Saily J, Koivisto P, Huhtinen I, Dufva T, Rautiainen A, Toivanen H, Nummila K (2013) Road surface condition detection using 24 GHz automotive radar technology, radar symposium (IRS). In: Proceedings IEEE applied electronics conference, vol 2, no 19–21, pp 702–707
8. Viikari V, Varpula T, Kantanen M (2009) Road-condition recognition using 24-GHz automotive radar. IEEE Trans Intell Transp Syst 10(4):639–648
9. Raj A, Krishna D, Hari Priya R, Kumar S, Niranjani Devi S (2012) Vision based road surface detection for automotive systems. In: Proceedings IEEE applied electronics conference, pp 223–228
10. Kees R, Detlefsen J (1994) Road surface classification by using a polarimetric coherent radar module at millimetre waves. In: Proceedings. IEEE national telesystems conference, pp 95–98
11. Kim HS (2001) Road surface sensing device. Korean patent KR 2001:047234
12. Childers Donald G, Skinner David P, Kemerait RC (1977) Cepstrum: a guide to processing. Proc IEEE 65(10):1–16

Rough Set-Based Analysis of Multilayer Stack Parameter on QoS of MANET

N. Pathviraj and Santosh L. Deshpande

Abstract The multi-constrained quality of service (QoS) routing in MANET considers multiple QoS parameters to route the packet from source to destination. Taking decision on which criteria different QoS parameters have to be taken into consideration is a critical task in dynamic network scenarios like MANET. The network scenario and stack parameters are aligned with different layers like application, MAC, physical change hand in hand. The set of these stack parameters has a significant impact on different QoS parameters, and trade-off point between these stack parameters is identified to estimate the significance of stack parameters in QoS parameters of the MANET. The rough set theory is used to identify trade-off point to estimate the significance of the stack parameter on different QoS parameters and QoS parameters to be considered in routing. The rough set decision rules are filtered out based on probabilistic properties like strength, certainty, and coverage of the decision rules. The positive dependency between conditions and decisions of selected decision rules is confirmed with certainty and coverage of relationships. The Rough Set Exploration System (RSES) is used for extensive analysis of QoS in data delivery under seven different stack parameter configurations like packet inter-arrival time, payload, and retry delay. The different stages of RSES like data exploration, discretization, reduction, and decision rules are applied to QoS values of different routes. The empirical model of stack parameters is derived for multi-parameter optimization to get best-effort QoS in all routes of the MAENT. The decision rules are selected based on strength and certainty, and coverage of decision rules will be used in the routing algorithm to get best QoS in MANET routing.

N. Pathviraj (✉) · S. L. Deshpande
Department of PG Studies, Visvesvaraya Technological University,
Jnana Sangama, Belagavi, India
e-mail: praj151986@gmail.com

S. L. Deshpande
e-mail: sld@vtu.ac.in

© Springer Nature Singapore Pte Ltd. 2019
N. Chaki et al. (eds.), *Proceedings of International Conference on Computational Intelligence and Data Engineering*, Lecture Notes on Data Engineering and Communications Technologies 28, https://doi.org/10.1007/978-981-13-6459-4_22

1 Introduction

The mobile ad hoc network (MANET) is dynamic and on the fly setup without any involvement of access point, router, and any fixed infrastructure. The dynamism of the MANET differentiates this network from other variants of networks. Wireless connection among the mobile nodes leads to data loss because of environmental inference, fading, and obstacles. The devices used in this class of network have limited transmission range; this leads to multi-hop communication. The devices used in remote location relay on the battery power, causes failure of mobile nodes because of power depletion. The topology of nodes will change as nodes change its position. These issues make the MANET more unreliable network and challenging to provide good quality of service (QoS) over different network paths [1].

The key issue in MANET is to improve the performance by QoS provisioning for different applications. The quality of service provisioning in packet delivery ensures reliable delivery along the route from source to destination. The network conditions like node mobility, node and link failures, inference, and energy have high impact on ensuring QoS for the route [2]. There are three fundamental challenges in forwarding the packets in QoS-complained routes. First, the QoS metric of the entire path is computed link by link based on route request packet being forwarded. Second, broadcast-based routing algorithms do not explore all possible routes. Third, bandwidth computation mechanisms on a link are based on other parameters like queue length and channel access history.

The traditional approach focuses on one or two fixed QoS metrics and tries to route packets by considering these QoS aspects. This may lead to negative impact on other QoS parameters just because network scenario keeps changing. Especially in the large-scale MANET scenario, it is so difficult to come up with the route that is aligned with multiple QoS constraints. So, it is proven that establishing route with multiple QoS aspects is NP-complete problem [3]. The dynamic decision-making approach will be an effective theory to solve the problem of multi-constrained QoS routing. The dynamic multi-constrained decision making will provision the QoS on per-flow basis depending on the protocol stack parameters. Multi-constrained decision making recommends the rule set based on the evaluation of multiple objects from different points of view.

The better quality of service can be provided by optimizing parameters in different layers of the protocol stack. The QoS performance metrics are measured and optimized based on the effects of different stack parameters. The joint effects of multiple stack parameters on the different QoS aspects will vary because of values at different layers of the protocol stack. So fine tuning parameter guidelines to finalise the trade-off point that will yield better QoS [4]. The fine-tuning not only gives the joint impacts but also provides background to find the right parameter configuration for a broad range of link quality to achieve the optimization goal of multiple QoS metrics in dynamic channel conditions.

The rough set theory (RST) is incorporated to derive the efficient dynamic routing decision rule with minimum number of QoS attributes. The different stages of RST like discretization, reduction, decision rule are applied to original routing table, and finally decision rule will be derived. The probabilistic properties of the decision rule will be calculated to get to know the strength, consistency, and coverage of the decision rule [5]. The selected decision rules based on probabilistic properties are included in the routing algorithm to get the better QoS.

The rest of the paper is organized as follows. In Sect. 2, network model with QoS is discussed. In Sect. 3, shortcomings of different existing QoS models are discussed. Section 4 covers the rough set theory introduction. In Sect. 5, trade-off with stack parameter for a QoS parameter is analyzed. Section 5.1 covers the trade-off between QoS parameter discussed along with different stages of RST, and probabilistic properties of the decision rule are calculated. In Sect. 6, article is concluded.

2 Network Model with QoS

A network can be depleted as a weighted digraph $G = (V, E)$, where V represents the set of nodes and E represents the set of communication links between the nodes. $|V|$ and $|E|$ represent the total number of nodes and links in the network, respectively. For $\forall\ v_i \in V$, $\forall\ v_j \in V$, and $v_i \neq v_j$ in $G(V, E)$, (i, j) represents the bidirectional path between v_i and v_j in MANET. The QoS model takes into consideration all the QoS parameters in the communication link to be followed from source to destination [6]. Consider $P = (V_s, V_d)$ represents the communication link between source and destination, where $V_s \in V$ and $V_d \in (V - \{V_s\})$.

The weight of the entire path from source to destination is a consolidated effect of all the QoS parameters. Consider multi-hop path from source s to destination d consists of intermediate nodes $u_1, u_2 ... u_j$. Let QoS (s, d) represent the QoS weight of the entire path from source to destination based on the QoS values of each link that are part of the path. The QoS weight findings are broadly classified into three broad categories to simplify techniques to calculate QoS of the entire path:

Additive: A QoS weight of the entire path is the summation of all the links participating in the communication.

$$QoS(s, d) = QoS(s, u_1) + \cdots + QoS(u_i, d) \tag{1}$$

For example, hop count and delay are the additive QoS parameters.

Multiplicative: A QoS weight is the multiplication of all the links participating in the communication.

$$QoS(s, d) = QoS(s, u_1) \times \cdots \times QoS(u_i, d) \tag{2}$$

For example, probability of packet loss and reliability are the multiplicative QoS parameters.

Concave: A QoS weight of the entire path is the comparative or minimum of the links participating in the communication.

$$QoS(s, d) = \min\{QoS(s, u_1), \ldots, QoS(u_i, d)\} \tag{3}$$

For example, bandwidth is the concave QoS parameter.

3 Shortcomings of Existing QoS Models

The different contexts of wireless ad hoc networks compared to fixed networks necessitate different techniques in providing QoS in the network. Several QoS architectures have been proposed for the ad hoc network independent of the fixed network. They are classified into three broad categories, namely class-based, flow-based, and hybrid QoS models. The QoS models [7] have been standardized by the Internet Engineering Task Force (IETF), namely integrated service (IntServ), differentiated service (DiffServ), and flexible QoS model for MANET (FQMM). The following sections describe these three QoS models.

3.1 *IntServ in MANETS*

The integrated service (IntServ) model is based on the maintaining flow-specific information in every intermediate router. A flow is a session between the source and the destination at the application layer. Here, flow-specific state should maintain QoS information like bandwidth requirement, energy, and delay bound of each flow. The IntServ model delivers controlled load service and guaranteed service by reserving the resources before the packet transmission [8]. The applications always initiate transmission only after allocating or reserving the required resources.

The maintenance of per-flow information in all intermediate nodes leads to processing and storage overhead. But this kind of scalability issues may not occur in present generation MANET by considering the bandwidth of links, limited number of hops, and less number of flows. The signaling protocols adopt complex handshaking mechanism, lead to consume lot of bandwidth, and may become too slow. IntServ model processes the admission control, classification, and scheduling in each and every limited resource mobile nodes. This will be a burden on the network scenarios like MANET.

3.2 DiffServ in MANETS

The differentiated service (DiffServ) model negates the problems of implementing and deploying IntServ by avoiding maintaining per-flow state information. Here, limited numbers of generalized classes are maintained to solve the scalability issues of IntServ. The interior nodes have to forward the packets by considering limited number of generalized classes; this makes the interior nodes' job simpler [9]. So, DiffServ will become the lightweight process for the intermediate nodes and lead to consider this model as a possible solution to QoS problem.

The relative priority approach for generalized class of QoS parameter is incorporated in DiffServ model. This will lead to soften QoS guarantee in terms of not offering best QoS. The service level agreement (SLA) is based on the concept of contract between customer and service provider. The DiffServ works based on the SLA and tries to assign sufficient resources as per SLA. In ad hoc networks, there is no concept of service provider and assuring the resources is also not possible, so SLA is not feasible in MANET. The main advantage of DiffServ is generalization of traffic among different classes at the boundary node itself. Since every node is potential sender and receiver, it is very difficult to identify boundary nodes.

3.3 Flexible QoS Model for MANET

The FQMM is the hybrid QoS model which incorporates benefits like flow granularity from IntServ model and service differentiation from DiffServ model. A traffic conditioner of the IntServ is placed only in entry nodes where packet originates and the interior nodes forward the packets on the aggregated per class basis like in DiffServ. Since FQMM incorporates from both IntServ and DiffServ, handling issues with respect to plug and play of concepts is the critical task [10]. Implementing aggregated per class flow in each and every internal node and dynamically negotiated traffic flow is the difficult job.

4 Rough Set Theory

The rough set theory (RST) is the right tool in dynamic decision making based on a wide variety of data set coming from protocol stack parameters. The RST is most suitable technique to deal with both quantitative and qualitative data sets, and filtering out data inconsistencies also made it simpler. It is possible to extract relevance of the particular attribute, lower and upper approximation of the data set into decision making, and prepare the ground for extracting knowledge in the form of decision rules. These decision rules are type of logical statement; it can be directly used in packet routing from source to destination [11]. The RST is extracting the decision

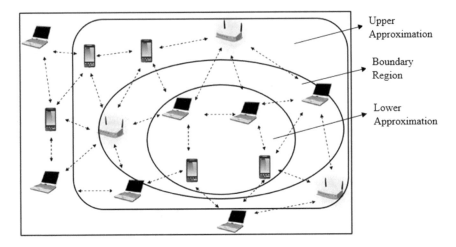

Fig. 1 Levels of approximation in rough set theory

rules from the data set without losing granularity of each and every conditional attribute.

The RST is an excellent mathematical tool to build the decision model from the large set of inconsistent data with multiple attributes. Extracting decision model in terms of rules from inconsistent data is a crucial task without considering any additional information on data [7]. This concept can be considered as another approach to deal with vagueness, i. e, imprecise in RST unlike precise in fuzzy set theory [12]. Suppose for a given QoS parameter of the different route is considered as a set of objects U called universe and an equivalence relation $R \in U \times U$. Let X be a subset of U (Fig. 1).

The characterization of set X with respect to R with the three basic concepts of rough set theory is:

The lower approximation of set X with respect to R is the set of all QoS routes, which may be certainly classified as X aligned with R.

The upper approximation of set X with respect to R is the set of all QoS routes, which may be possibly classified as X aligned with R.

The boundary region of set X with respect to R is the set of all QoS routes, which may be classified neither as X nor as not $- X$ aligned with R.

5   QoS Parameter Optimization

The QoS parameter of MANET is contributed by multiple stack parameters; these parameters are from application, MAC, and physical layer. The seven stack parameters along with which layer they belong to are mentioned in Table 1. Here, explorative

Table 1 Stack parameter description at different layers

Layers	Parameters	Values	Description
App	Tpit: Packet inter-arrival time (ms)	10,15, 20, 25, 30, 35, 40, 50	Tpit is inversely proportional to packet arrival time
	L_d: Packet payload size (bytes)	20, 35, 50, 65, 80, 95, 110	Directly proportional impact of l_D on goodput, delay
MAC	Qmax: Maximum queue size (packets)	1, 30, 60	The values represent no queue, and medium and large queue
	Nmaxtries: Maximum number of transmission	1, 3, 5	Maximum number of packet retransmission in packet lost
	Dretry: Retry delay (ms)	30, 60	Time gap between retry
Phy	Ptx: Transmission power level	3, 7, 11, 15, 19, 23, 27, 31	Transmission power level to transfer the packet
	d: Distance between nodes (meter)	10, 15, 20, 25, 30, 35	Different distances between nodes in network scenario

study is conducted to analyze the performance trade-off between QoS parameters and seven stack parameters. This study gives potential gain in understanding which network scenario yields better QoS and not [13]. It will be easier for decision makers to come up with new QoS model that consumes lesser resources and gives better QoS.

The stack parameter configuration is the plug and play of seven different combinations for different stack parameter values. Overall 4500 packets are communicated between sender and receiver under different combinations of stack parameter values. All experiments are conducted based on the distance, for different combinations of remaining six parameters [13]. Here, both sender and receiver maintain packet information that contains received signal strength indication (RSSI), link quality indicator (LQI), actual queue size, transmission number, and time of receiving. An analysis has been conducted on the measured data for different stack configurations and discussed outcomes from packet error rate, energy efficiency, goodput, and delay.

The Rough Set Exploration System (RSES) is used for data exploration, discretization, reduction, and decision making [14]. It is an open-source tool that helps in understanding tabular data set generated from stack parameter configuration and makes decision with respect to reason for variation in different QoS parameters [15]. In order to ensure, good decision has been made accuracy and coverage of the decision will be displayed for all QoS parameters. For analyzing stack parameters, trade-off in different QoS parameters using RSES follows below-mentioned steps:

1. Load the sample data set into the RSES.

2. Load the cut set rules into RSES.
3. Generate the discretization table.
4. Derive the decision rule.
5. Generate statistics for rule set.

5.1 Trade-off Between Stack Parameters in QoS Parameter

5.1.1 Packet Error Rate (PER)

PER is the ratio of the number of unacknowledged packets sent to the total number of packets sent with acknowledgment. The packet transfer is more related to physical layer aspects; so, study focuses on packet payload size l_D and SNR. The rule set in Table 2 shows the correlation of PER with packet payload size and SNR. Interesting observation made here is PER is inversely proportional to the SNR; means as SNR increases PER decreases. The decision rules were able to justify PER increases with packet payload size (Tables 3 and 4).

5.1.2 Energy

The energy efficiency indicates energy consumption for the per packet data transfer in the respective nodes. Observation is made on energy choosing the optimized stack parameter configuration. The impact of transmission power, packet payload size, and SNR on the energy consumption is shown in rule set of Table 5. The energy consumption stabilizes when there is low noise in the physical medium, so there

Table 2 PER decision rules

(1–3)	Match	Decision rules
1	7	(SNR=6)$\Longrightarrow$(PER={(0.25,Inf)[7]})
2	7	(SNR=16)$\Longrightarrow$(PER={(-Inf,0.25)[7]})
3	7	(SNR=26)$\Longrightarrow$(PER={(0.25,Inf)[7]})

Table 3 Decision rule accuracy

	(0.25, Inf)	(-Inf, 0.25)	No. of objects	Accuracy	Coverage
(0.25, Inf)	7	0	7	1	1
(-Inf, 0.25)	0	14	14	1	1
True positive	1	1			

Total number of tested objects: 21
Total accuracy: 1
Total coverage: 1

Table 4 PER discretized values

21/3	SNR	L_D	PER	21/3	SNR	L_D	PER
0:1	6	20	(0.25, Inf)	0:11	16	65	(-Inf, 0.25)
0:2	6	35	(0.25, Inf)	0:12	16	80	(-Inf, 0.25)
0:3	6	50	(0.25, Inf)	0:13	16	95	(-Inf, 0.25)
0:4	6	65	(0.25, Inf)	0:14	16	110	(-Inf, 0.25)
0:5	6	80	(0.25, Inf)	0:15	26	20	(-Inf, 0.25)
0:6	6	95	(0.25, Inf)	0:16	26	35	(-Inf, 0.25)
0:7	6	110	(0.25, Inf)	0:17	26	50	(-Inf, 0.25)
0:8	16	20	(-Inf, 0.25)	0:18	26	65	(-Inf, 0.25)
0:9	16	35	(-Inf, 0.25)	0:19	26	80	(-Inf,0.25)
0:10	16	50	(-Inf, 0.25)	0:20	26	95	(-Inf, 0.25)
				0:21	26	110	(-Inf, 0.25)

Table 5 Energy decision rules

(1–6)	Match	Decision rules
1	9	(Tx=3)$\Longrightarrow$(Ueng={(0.12, Inf)[9]})
2	6	(SNR=5)$\Longrightarrow$(Ueng={(0.12, Inf)[6]})
3	3	(Tx=11)&(SNR=17)$\Longrightarrow$(Ueng={(-Inf, 0.12)[3]})
4	2	(SNR=11)&(l_D=110)$\Longrightarrow$(Ueng={(0.12, Inf)[2]})
5	1	(Tx=11)&(SNR=11)&(l_D=20)$\Longrightarrow$(Ueng={(-Inf, 0.12)[1]})
6	1	(Tx=11)&(SNR=11)&(l_D=70)$\Longrightarrow$(Ueng={(-Inf, 0.12)[1]})

Table 6 Decision rule accuracy

	(0.25, Inf)	(-Inf, 0.25)	No. of objects	Accuracy	Coverage
(0.12, Inf)	13	0	13	1	1
(-Inf, 0.12)	0	5	5	1	1
True positive	1	1			

Total number of tested objects: 18
Total accuracy: 1
Total coverage: 1

is less chance for loss of packets. It has been concluded in the decision rule that as packet payload size increases there is a gradual hike in the energy consumption per packet. The energy consumption will be optimized if the noise in the physical medium is high (SNR = 5) with less or medium-sized packets and with the increasing SNR values for high and medium-sized packets (Tables 6 and 7).

Table 7 Energy discretized values

18/4	Tx	SNR	l_D	Ueng	18/4	Tx	SNR	l_D	Ueng
0:1	3	5	20	(0.25, Inf)	0:10	11	5	20	(-Inf, 0.25)
0:2	3	5	70	(0.25, Inf)	0:11	11	5	70	(-Inf, 0.25)
0:3	3	5	110	(0.25, Inf)	0:12	11	5	110	(-Inf, 0.25)
0:4	3	11	20	(0.25,Inf)	0:13	11	11	20	(-Inf, 0.25)
0:5	3	11	70	(0.25, Inf)	0:14	11	11	70	(-Inf, 0.25)
0:6	3	11	110	(0.25, Inf)	0:15	11	11	110	(-Inf, 0.25)
0:7	3	17	20	(0.25, Inf)	0:16	11	17	20	(-Inf, 0.25)
0:8	3	17	70	(-Inf, 0.25)	0:17	11	17	70	(-Inf, 0.25)
0:9	3	17	110	(-Inf, 0.25)	0:18	11	17	110	(-Inf, 0.25)

Table 8 Goodput decision rules

(1-5)	Match	Decision rules
1	3	$(SNR=5) \Longrightarrow (goodput=\{(-Inf, 25.0)[3]\})$
2	3	$(l_D=20) \Longrightarrow (goodput=\{(-Inf, 25.0)[3]\})$
3	1	$(SNR=11)\&(l_D=70) \Longrightarrow (goodput=\{(-Inf, 25.0)[1]\})$
4	1	$(SNR=11)\&(l_D=110) \Longrightarrow (goodput=\{(25.0, Inf)[1]\})$
5	1	$(SNR=17)\&(l_D=70) \Longrightarrow (goodput=\{(25.0, Inf)[1]\})$

5.1.3 Goodput

The goodput is the application level throughput; that is, the number of data in bits is transferred per unit time. This observation is made for choosing the optimized stack parameters to get better goodput under different link conditions. The goodput increases with reducing noise in the physical medium, and it is also having joint impact of the packet payload size as shown in rule set of Table 8. Here, SNR intern influences the transmission power required, noise in the physical medium is high leads to transmission power required will also be high. If SNR values are in 5 db, impacts of the packet payload are not seen in the goodput but at the same time as SNR leads to 17 db impact is clearly visible. The goodput increases with increasing packet payload size and with higher SNR values. The goodput variation is minimal with respect to lower SNR values because of high noise in the physical medium.

5.1.4 Delay

The delay perceived in the transfer of packets from source to destination consists of queuing delay and service time delay. The delay is jointly impacted by packet inter-arrival time, packet payload size, and SNR as shown in rule set of Table 12. When all the nodes are having low SNR with the effect of noise in the physical medium, there

Table 9 Decision rule accuracy

	(0.25, Inf)	(-Inf, 0.25)	No. of objects	Accuracy	Coverage
(25.0, Inf)	6	0	6	1	1
(-Inf, 25.0)	0	2	2	1	1
True positive	1	1			

Total number of tested objects: 8
Total accuracy: 1
Total coverage: 1

Table 10 Goodput discretized values

8/3	SNR	I_D	Goodput
0:1	5	20	(-Inf, 25.0)
0:2	5	70	(-Inf, 25.0)
0:3	5	110	(-Inf, 25.0)
0:4	11	20	(-Inf, 25.0)
0:5	11	70	(-Inf, 25.0)
0:6	11	110	(25.0, Inf)
0:7	17	20	(-Inf, 25.0)
0:8	17	70	(25.0, Inf)

Table 11 Decision rule accuracy

	(25.0, Inf)	(-Inf, 25.0)	No. of objects	Accuracy	Coverage
(25.0, Inf)	2	0	2	1	1
(-Inf, 25.0)	0	10	10	1	1
True positive	1	1			

Total number of tested objects: 21
Total accuracy: 1
Total coverage: 1

Table 12 Delay decision rules

(1–4)	Match	Decision rules
1	6	$(I_D=20) \Longrightarrow (\text{Delay}=\{(-\text{Inf}, 25.0)[6]\})$
2	4	$(SNR=10) \Longrightarrow (\text{Delay}=\{(-\text{Inf},25.0)[4]\})$
3	4	$(SNR=30) \Longrightarrow (\text{Delay}=\{(-\text{Inf}, 25.0)[4]\})$

will be a significant hike in the queuing delay. The packet payload size and packet inter-arrival time have positive impact or are directly proportional to the delay for the same SNR values. Both of these parameter configurations have less impact when it is compared with respect to SNR (Tables 9, 10, 11, and 13).

Table 13 Delay discretized values

12/4	P_{int}	l_D	SNR	Delay
0:1	10	110	5	(25.0, Inf)
0:2	10	110	10	(-Inf, 25.0)
0:3	10	110	30	(-Inf, 25.0)
0:4	10	20	5	(-Inf, 25.0)
0:5	10	20	10	(-Inf, 25.0)
0:6	10	20	30	(-Inf, 25.0)
0:7	30	110	5	(25.0, Inf)
0:8	30	110	10	(-Inf, 25.0)
0:9	30	110	30	(-Inf, 25.0)
0:10	30	20	5	(-Inf, 25.0)
0:11	30	20	10	(-Inf, 25.0)
0:12	30	20	30	(-Inf, 25.0)

5.1.5 Analysis of Trade-off Between Stack Parameters

Multi-constrained QoS routing in MANET is the joint consideration of multiple QoS aspects. Each and every QoS parameter is impacted by set of stack parameters, to enhance the overall QoS support to get best-effort service; threshold level of such stack parameters needs to be identified. From the rough set theory findings, it has been observed that energy consumption will be less if Tx and SNR are high, PER will be less if SNR value is high, goodput will be high if SNR and lD are high, and finally delay will be less if SNR and lD are high. Multi-constrained QoS-optimized packet delivery happens only with max (SNR), max (TX), and max (lD). So, multi-parameter optimization is current requirement to get the best-effort QoS. The empirical model to get the good QoS is:

$$\boxed{\textbf{QoS (SNR, T}_x, \textbf{l}_D\textbf{)}} \qquad (4)$$

6 Conclusion

The extensive analysis of packet delivery information enables us to come to an understanding on impact of key stack parameters on different QoS parameters. The trade-off point between the key stack parameters is analyzed with the help of rough set theory to improve the QoS aspects of packet delivery. The RSES is used for analysis of QoS routes in different stages like discretization, reduction, decision rules of the RST. The empirical model of stack parameters consists of SNR, transmission power and packet payload size is derived from rough set analysis. Trade-off point between these stack parameters contributes to the better QoS-constrained packet delivery. This multi-constrained and multi-objective analysis through RST is helpful for the decision makers to take dynamic decision on QoS for dynamic network scenarios like MANET.

References

1. Lahyani I, Khabou N, Jmaiel M (2012) QoS monitoring and analysis approach for publish/subscribe systems deployed on MANET. In: 20th Euromicro international conference on parallel, distributed and network-based processing, pp 120–124
2. Rao M, Singh N (2015) Simulation of various QoS parameters in a high density Manet setup using AODV nthBR protocol for multimedia transmission, data transmission and under congestion scenario. ICTACT J Commun Technol 06(03):1155–1159
3. Johari L, Mishra RK (2016) A review on the recent quality-of-services (QoS) issues in MANET. Int J Comput Appl (0975 – 8887) 139(5):27–33
4. Mohapatra P, Li Jian, Gui Chao (2003) QoS in mobile ad hoc networks. IEEE Wirel Commun 10(3):44–52
5. Pawlak Z (1982) Rough sets. Int J Comput Inform Sci 11(5):341–356
6. Polkowski L (2002) Rough sets—mathematical foundations. In: Advances in soft computing, Physica Verlag, Springer Verlag Company, 1-534
7. Pawlak Z (2000) Rough sets and decision algorithms. Rough sets and current trends in computing, LNAI, pp 30–45
8. Braden R, Clark D, Shenker S (1994) Integrated services in the Internet architecture: an overview. Technical Report 1633
9. Black D (2000) Differentiated services and tunnels. RFC2983
10. Xio H, Seah WKG, Lo A, Chua KC (2000) A flexible quality of service model for mobile ad-hoc networks. IEE VTC2000-spring, Tokyo, Japan
11. Pawlak Z (1991) Theoretical aspects of reasoning about data. Kluwer Academic Publishers
12. Pawlak Z (1998) An inquiry into anatomy of conflicts. J Inform Sci 109:65–68
13. Fu S, Zhang Y, Jiang Y, Hu C, Shih C-Y, Marrón PJ (2015) Experimental study for multi-layer parameter configuration of WSN links. In: IEEE 35th international conference on distributed computing systems, pp 369–378
14. The RSES. http://logic.mimuw.edu.pl/~rses
15. Lian J, Li L, Zhu X (2007) A multi-constraint QoS routing protocol with route-request selection based on mobile predicting in MANET. In: International conference on computational intelligence and security workshops, pp 342–345

A Dynamic Object Detection In Real-World Scenarios

Kausar Hena, J. Amudha and R. Aarthi

Abstract The object recognition is one of the most challenging tasks in computer vision, especially in the case of real-time robotic object recognition scenes where it is difficult to predefine an object and its location. To address this challenge, we propose an object detection method that can be adaptive to learn objects independent of the environment, by enhancing the relevant features of the object and by suppressing the other irrelevant feature. The proposed method has been modeled to learn the association of features from the given training dataset. Using dynamic evolution of neuro-fuzzy inference system (DENFIS) model has been used to generate number of rules from the cluster formed from the dataset. The validation of the model has been carried on various datasets created from the real-world scenario. The system is capable of locating the target regardless of scale, illumination variance, and background.

Keywords Computer vision · Fuzzy system · Region of interest

1 Introduction

Object detection is a technique widely used in computer vision to find an object or multiple objects in images. There are two different approach, in the first approach, the object detection starts with the image segmentation, then the features are extracted and classified on the segmented regions. However, the second approach works on the principle of human visual system, where the interested regions are identified specific to the object of search in the complex environment and ignores the irrelevant

K. Hena · J. Amudha (✉)
Department of Computer Science and Engineering,
Amrita School of Engineering, Amrita Vishwa Vidyapeetham, Bengaluru, India
e-mail: j_amudha@blr.amrita.edu

R. Aarthi
Department of Computer Science and Engineering, Amrita School of Engineering,
Amrita Vishwa Vidyapeetham, Coimbatore, India

© Springer Nature Singapore Pte Ltd. 2019
N. Chaki et al. (eds.), *Proceedings of International Conference on Computational Intelligence and Data Engineering*, Lecture Notes on Data Engineering and Communications Technologies 28, https://doi.org/10.1007/978-981-13-6459-4_23

information/regions. Generally, human use cues like location, color, shape, and size of the environment to help humans to locate the objects. These cues are highly correlated with the position of camera, lighting and illumination and background which when modeled as vision system makes the design quite complex. There are quite few object recognition features like SIFT, BoW [1], and SURF [2] which are known as invariant feature derived to address these issues. Few methods handle colors in transformed space like Lab, Ycbcr [3] color space to overcome the problem of color dependencies. Classification methods like KNN, ADABOOST, SVM [4], and deep learning [5] help to identify the feature subset and to classify the object.

Object recognition implies the computation of set of target features at first step and their combinations in the further steps. In early stages of object recognition, template-based method gives an excellent performance for a single-object category detection, such as faces, cars, and pedestrians [6–8]. Histogram-based descriptors [2, 9] are developed to handle multiple categorizations and object transformation. The SIFT [1] and SURF features [2] are shown better way of deriving feature descriptors that are independent of image transformations. Part-based and appearance-based are other variations of object detection that take shape as a major cue to identify objects. Few advanced methods use contextual knowledge of the object in a scene to detect the object [10, 11]. For example, to detect the pedestrians in the given scene, search can be approximated to road or walking pathway region eliminating the sky. Many applications in the computer vision like video summarization [12], compression [13], robotic navigation and localization [14, 15] are addressed using object recognition and detection methods. In robotic applications, models are developed based on teach and replay scenario. For example, in navigation scenario the robots are manually shown the important landmark for several number of times to learn the navigation path [16]. In case of object detection and learning, the objects were taught to machine several number of times and increasing the object one by one [17]. The real-time system requires adaptive learning behavior to mimic the behavior of humans. Though the recent system solves many real-world problem, they differ from the way human remembers the objects. The human vision system is intelligent in extracting the dominating cues and learning their association in a simple manner. Hence, the objective of the work is to automatically select the optimal cues and their relations. One of the major problems is to distinguish the feature that are different from other and binding their association for a single object. Fuzzy system is one of the methods that solves the problem by forming the rules association. The fuzzy rules are effectively used to identify the distinct object in the images [18]. The main contribution of the dynamic object detection system has been toward developing a dynamic object detection system which can learn the objects of the scene on the fly, based on the surrounding environment where it has to be deployed for use. The DENFIS [19, 20] approach is used to modeling the relationship of object with its features. Effectiveness of the method is analyzed by increasing the object samples.

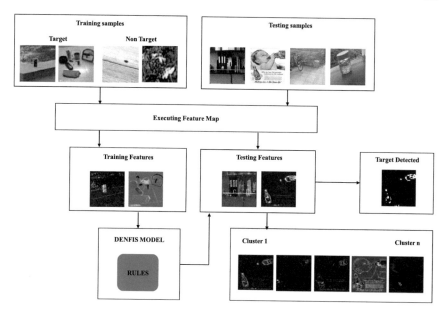

Fig. 1 Proposed architecture

2 Dynamic Object Detection System

The dynamic object detection system has different stages as described in Fig. 1. The aim of this work is to design a target detection application which will display the target region by suppressing the background information. The detection system should be able to detect the target irrespective of different background and different context. Figure 1 describes dynamic object detection system architecture.

2.1 Image Dataset

We have created a dataset of the objects that are used in day-to-day life. The images are self-created or downloaded from Web to form the image database. Presence of object in the images is selected from different scenarios, location, and distance. The image may consist of multiple objects. The training samples are created from image dataset by selecting the location of target object. Remaining regions are selected for background.

2.2 *Feature Extraction*

The common way to represent the object is based on its color, shape, and brightness. The mean values feature (Fi) from $i = 1, 2 \ldots 13$ where extracted from training samples. Color features (F1…F4) are separated into red, green, blue, and yellow. In addition to that saliency value of object is calculated in color space double opponency RG, GR, BY, and YB (F5…F8) by center surround differences across the normalized color channels [13]. RG and BY color opponency feature maps are computed at the center by calculating (red–green), and surrounds are subtracted by (green–red). The orientation maps obtained from Gabor filter 00° , 45° , 90°, 135° (F9…F12) and intensity (F13) are computed using oriented pyramids as in VOCUS [21]. After the features have been extracted, the DENFIS model will be designed for the cluster and rules generation for target object detection.

2.3 *Design Using DENFIS*

The main advantage of DENFIS model is an online evolving clustering method. Clusters are defined by the center (Ci) and radius (Ri), and the initially values are set to zero. Out of the total samples (T), the clustering process starts with assigning the first training sample initial cluster. For remaining samples (m) are merged with the closet distance cluster $M < Rj$, where Rj is the cluster radius of the current cluster center (Cj). If cluster radius, Ri, reaches the threshold value (Disthr/2) create new cluster with the sample. Otherwise, the cluster center and radius are updated based on added sample. Triangular membership functions are used for DENFIS model for rule generation which are defined by parameters x, y, and z, where y is the value of the center, $x = y - t * Disthr$, and $z = y + t * Disthr$, t is the threshold value which is 1.2, Disthr is the clustering parameter used as 0.9, and the membership value of sample (m) for the cluster is evaluated using Eq. 1

$$mf(m, x, y, z) = \max(\min((m - x)/(y - x), (z - m)/(z - y)), 0) \qquad (1)$$

When each cluster center and radius forms the membership methods in the feature space. The dataset has been fed to the DENFIS model to generate the clusters and rules. Table 1 shows the generated cluster structure, and totally four rules were generated as shown in Table 2. The test sample received the calculated degree of membership for each of the clusters. Table 1 displays the cluster structure of coke dataset, trained with 50 samples, and each sample has different background and tested on some of the samples which is mentioned in Table 1.

Table 1 shows cluster structure; each cluster is represented by center, radius, and members. The consequent values of the cluster give the knowledge about different classes available in training dataset. In this case, it is between target versus non-target samples.

Table 1 Cluster structure

Name	Center	Radius	Members	Out members	Consequent
Cluster 1	1 × 13 double	0.786383383405926	19 × 13 double	19 × 1 double	2
Cluster 2	1 × 13 double	0.886440164423723	7 × 13 double	7 × 1 double	1
Cluster 3	1 × 13 double	0.843648079633757	5 × 13 double	5 × 1 double	1
Cluster 4	1 × 13 double	0.570563004570589	3 × 13 double	3 × 1 double	1

Table 2 Rules

1	If (in1 is in1cluster1) and ... and (in13 is in13cluster1) then (out1 is out1cluster1) (1)[22]
2	If (in1 is in1cluster2) and ... and (in13 is in13cluster2) then (out1 is out1cluster2) (1)
3	If (in1 is in1cluster3) and ... and (in13 is in13cluster3) then (out1 is out1cluster3) (1)
4	If (in1 is in1cluster4) and ... and (in13 is in13cluster4) then (out1 is out1cluster4) (1)

Similarly, Table 2 discusses the rules generated by DENFIS model, which interpret the clusters belonging to features. First, the clusters and membership degree have been calculated for each feature. After that the rules will be generated for the training set. The knowledge acquired by the rule generated by DENFIS model has been used on the testing samples. In testing phase, the belonginess of the sample to each cluster is calculated. The membership belonging of each feature vector to clusters has been evaluated in the fuzzy system, and the clusters analysis has been done on resulting samples to retrieve the target object from testing samples.

3 Experimental Results

The system is implemented using MATLAB, and its intermediate results are shown in Fig. 1. The training set has been created with image dataset with target and non-target region of interest. The number of samples has been trained with target as well as non-target objects. Features described in Sect. 2.2 are extracted from each samples of training set to form training features vector and rules generated by DENFIS model [19, 20]. Features have been extracted from testing samples, and the fuzzy system has been used to utilize the generated rules in the clusters. Table 3 infers information like the objects used for the experiment, the number of target and non-target sample for each object, details of the clusters, rules formed, description of all dataset, its training time and testing time in seconds. The sample of the objects has been created from indoor or outdoor office environment and some collected from Web for both target and non-target, the equal number of samples for target and non-target has been collected, and the size of each samples has been resized to 640 × 480. Creation of clusters and rules depends upon the samples.

Table 3 Samples

Object name	Sample number		Number of clusters	Number of rules	Source	Training time (seconds)	Testing time (seconds)	Description
	T	NT						
Coke bottle	50	50	5	5	Self-created	14	2	Coke bottle samples collected from indoor and outdoor. The objects in different context
Bike	40	40	7	7	Collected from web	46.8438	3.0625	Web dataset where object in same context
Box	15	15	4	4	Self-created	15.2813	1	The box object pick from same context, and the box object surrounded by different distracter object
Cup	35	35	6	6	Self-created	52.4375	1.04688	The cup object pick from same context, but the cup object surrounded by different distracter object
Headphone	25	25	4	4	Office indoor dataset	15.2813	1.10938	Office desk dataset of headphone as an object, and the other object on the desk becomes distracter object
Keyboard	25	25	5	5	Collected from web	19.1719	1.10938	Office desk dataset of keyboard as an object, and the other object on the desk becomes distracter object
Mobile	25	25	4	4	Collected from web	18.0313	0.96875	Office desk dataset of mobile object, and the other object on the desk becomes distracter object
Mouse	25	25	7	7	Office indoor dataset	13.2813	1	Office desk dataset of mouse as an object, and the other object on the desk becomes distracter object

Each rule has its own membership degree calculated from cluster center and radius using Eq. 1. The test sample is fed to the fuzzy inference system to identify the cluster to which it can be associated. Each cluster has its unique id to identify. Clusters' weight will be calculated from Eq. 1; the weights of the cluster which are greater or equal to 1 belong to target object and others belong to non-target objects. Figure 2 represents the target and non-target cluster representations for testing sample. Cluster representation of different samples has been shown in Fig. 2. In the first testing sample, the target is 7 Up bottle, sample contains three different types and colors of bottle. The rules generated by DENFIS model [Sect. 2.3] show that the weights of cluster 2, 3, 4, and 5 are less than 1 and only the cluster 1 have weight greater or equal to 1. Hence, cluster 1 7 Up bottle becomes the foreground object and rest of the clusters becomes the background. The column 7 shows the segmentation result of test samples, and last column contains detected target of test system with red rectangle.

Table 4 shows the analysis that has been carried on, to find the target search by varying size of targets. The percentage area occupied by the target in the given image is calculated by,

$$\text{Area of target} = (\text{ size of target})/(\text{size of image})$$

As the area of target increases as shown in Table 4, the size of the object becoming larger in the image, Table 4 is divided in two categories depending upon the area of target as small and large. The first row gives detail for four smaller sample target

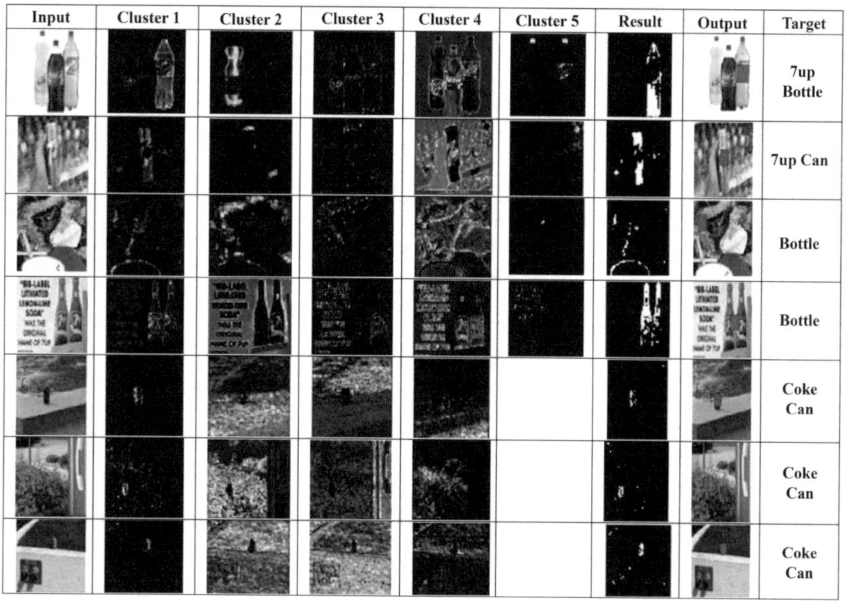

Input	Cluster 1	Cluster 2	Cluster 3	Cluster 4	Cluster 5	Result	Output	Target
								7up Bottle
								7up Can
								Bottle
								Bottle
								Coke Can
								Coke Can
								Coke Can

Fig. 2 Testing results

Table 4 Detected results for different size of objects

Area of target (in %)	Object name	# Test sample	#Detected	# Detected other objects
Small (1–30)	Box	15	15	0
	Cup	35	35	0
	Mouse	25	22	3
	Headphone	25	20	5
Large (31–90)	Bike	40	40	0
	Coke	50	50	0
	Mobile	25	25	0
	Keyboard	25	24	1

objects, followed by the number of actual object used for testing followed by actual number of target correctly detected and not detected. Similarly, the second row gives detail for objects size having larger area of target.

Table 5 shows the result of test samples where number of same target objects present in each sample. In Table 5, the coke bottle have 50 samples, out of which 46 samples contain only one bottle, 3 samples has two coke bottle, and 1 sample with 5 coke bottles. Similar scenario has been presented with bike, cup, box, and remaining targets which are grouped into others.

Table 6 shows the performance measures CC means correlation coefficient, AUC is area under curve, NSS is normalized scanpath saliency and SIMILARITY, of all objects used in this model for object detection, and these are compared with VOCUS model [21]. Twenty random images chosen from the eye tracking datasets of Bruce and Judd [23], and their corresponding eye fixation map is used to compare VOCUS and proposed model.

Table 5 Analysis based on # target present

Object name	# Test sample	#Target object present in test sample	#Detected	#Detected other objects
Coke bottle	46	1	50	0
	3	2		
	1	5		
Bike	46	1	40	0
	3	2		
	1	3		
Box	15	1	15	0
Cup	35	1	35	0
Others	100	1	68	9

Table 6 Classification accuracy

Model	CC	AUC	SIMILARITY	NSS
VOCUS	0.165	0.5939	0.3993	1.7745
PROPOSED	0.31	0.7	0.3	0.7
Coke bottle	0.6	0.7	0.3	0.7
Bike	0.3	1	0.2	0.2
Box	0.2	0.6	0.2	0.4
Cup	0.7	0.6	0.3	0.6
Headphone	0.3	0.7	0.3	0.7
Keyboard	0.2	0.6	0.2	0.5
Mobile	0.2	0.6	0.2	0.5
Mouse	0.2	0.6	0.2	0.5

4 Conclusion

An automated system has been generated, trained, and tested for the target search applications. Incorporating neural network to the existing target detection application will make it an adaptive model. Training and testing have been done on self-created and collected from Web image dataset. The proposed system has been able to detect the target whether the object present in same context or different context and will separate the foreground object out from background. The proposed system is detecting and categorizing the target object. The analysis has been done on multiple objects for evaluating the area of target of each object to check how much area of the test sample belongs to the target area correctly and number of objects detected correctly.

References

1. Sampath A, Sivaramakrishnan A, Narayan K, Aarthi R (2016) A study of household object recognition using SIFT-based bag-of-words dictionary and SVMs. In: Proceedings of the international conference on soft computing systems. Springer, New Delhi, pp 573–580
2. Lowe DG (1999) Object recognition from local scale-invariant features. In: The proceedings of the seventh IEEE international conference on computer vision, vol 2, IEEE, pp 1150–1157
3. Kaur A, Kranthi BV (2012) Comparison between YCbCr color space and CIELab color space for skin color segmentation. IJAIS 3(4):30–33
4. Alpaydin E (2014) Introduction to machine learning. MIT press, Dec 4
5. Li G, Yu Y (2016) Deep contrast learning for salient object detection. In Proceedings of the IEEE conference on computer vision and pattern recognition, pp 478–487
6. Schneiderman H, Kanade T (2000) A statistical method for 3D object detection applied to faces and cars. In: Proceedings IEEE conference on computer vision and pattern recognition, vol 1. IEEE, pp 746–751
7. Viola P, Jones M (2001) Robust real-time face detection. In: Proceedings. Eighth IEEE international conference on computer vision. ICCV, 2001. IEEE, p 747

8. Mohan A, Papageorgiou C, Poggio T (2001) Example-based object detection in images by components. IEEE Trans Pattern Anal Mach Intell 23(4):349–361
9. Serge B, Malik J, Puzicha J (2002) Shape matching and object recognition using shape contexts. IEEE Trans Pattern Anal Mach Intell 24(4):509–522
10. Torralba A (2003) A Contextual priming for object detection. Int J Comput Vis 53(2):169–191
11. Torralba A, Murphy K, Freeman W, Rubin M (2003) Context-based vision system for place and object recognition. In: IEEE international conference on computer vision (ICCV), pp 273–280
12. Ma Y, Hua X, Lu L, Zhang H (2005) A generic framework of user attention model and its application in video summarization. IEEE Trans Multimed 7(5):907–919
13. Itti L (2004) Automatic foveation for video compression using a neurobiological model of visual attention. IEEE Trans Image Process 13(10):1304–1318
14. Siagian C, Itti L (2009) Biologically inspired mobile robot vision localization. IEEE Trans Robot 25(4):861–873
15. Mertsching B, Bollmann M, Hoischen R, Schmalz S (1999) The neural active vision system. In: Jahne B, Haussecke H, Geissler P (eds) Handbook of computer vision and applications vol 3. Academic Press, pp 543–568
16. Siagian C, Chang CK, Itti L (2014) Autonomous mobile robot localization and navigation using a hierarchical map representation primarily guided by vision. J Field Robot 31(3):408–440
17. Pasquale G, Ciliberto C, Odone F, Rosasco L, Natale L (2015) Teaching iCub to recognize objects using deep convolution neural networks. In: Proceedings of The 4th workshop on machine learning for interactive systems at ICML 2015, PMLR, vol 43, pp 21–25
18. Lin W-S, Fang C-H (2006) Computational model of intention-oriented visual attention. In: IEEE international conference on systems, man, and cybernetics October pp 8–11
19. Kasabov NK, Song Q (2002) DENFIS: dynamic evolving neural-fuzzy inference system and its application for time-series prediction. IEEE Trans Fuzzy Syst 10(2):144–154
20. Amudha J, Radha D, Smitha S (2015) Analysis of fuzzy rule optimization models. Int J Eng Technol (IJET) 7:1564–1570
21. Frintrop S (2006) VOCUS.: a visual attention system for object detection and goal-directed search, vol 3899, Lecture Notes in Artificial Intelligence (LNAI). Springer, Heidelberg, Germany
22. Fadhilah R (2016) Fuzzy petri nets as a classification method for automatic speech intelligibility detection of children with speech impairments/Fadhilah Rosdi. Diss. University of Malaya
23. Bruce N, Tsotsos J (2006) Saliency based on information maximization. In: Advances in neural information processing systems, pp 155–162

Software Defect Prediction Using Principal Component Analysis and Naïve Bayes Algorithm

N. Dhamayanthi and B. Lavanya

Abstract How can I deliver defect-free software? Can I achieve more with less resources? How can I reduce time, effort, and cost involved in developing software? Software defect prediction is an important area of research which can significantly help the software development teams grappling with these questions in an effective way. A small increase in prediction accuracy will go a long way in helping software development teams improve their efficiency. In this paper, we have proposed a framework which uses PCA for dimensionality reduction and Naïve Bayes classification algorithm for building the prediction model. We have used seven projects from NASA Metrics Data Program for conducting experiments. We have seen an average increase of 10.3% in prediction accuracy when the learning algorithm is applied with the key features extracted from the datasets.

Keywords Software defect prediction · Fault proneness · Classification · Feature selection · Naïve Bayes classification algorithm · Principal component analysis · Software quality · Machine learning algorithms · Fault prediction · Dimensionality reduction · Data mining · Machine learning techniques · NASA Metrics Data Program · Stratified tenfold cross-validation · Reliable software · Prediction modeling

1 Introduction

No longer customer satisfaction is enough. It is all about experience. We need a high-quality software to create the experience required for building loyal customer base [1].

N. Dhamayanthi (✉) · B. Lavanya
Department of Computer Science, University of Madras, Chennai, India
e-mail: dhamayanthin@outlook.com

B. Lavanya
e-mail: lavanmu@gmail.com

© Springer Nature Singapore Pte Ltd. 2019
N. Chaki et al. (eds.), *Proceedings of International Conference on Computational Intelligence and Data Engineering*, Lecture Notes on Data Engineering and Communications Technologies 28, https://doi.org/10.1007/978-981-13-6459-4_24

As the expectation on quality of the software is increasing, there is a tremendous focus on the research on defect prediction [2]. Predicting the faulty modules will significantly help in deploying high-quality software.

While the research on software defect prediction has been going on for over three decades, leveraging machine learning for prediction has started gaining traction in recent years [3]. Hence, software development community is looking for more researchers to contribute to this area.

In this paper, we explore whether reducing features using PCA before applying Naïve Bayes classifier would improve the classification accuracy than using the full dataset.

Seven datasets from NASA MDP [4] PC1, PC2, PC3, CM1, KC1, KC3, and MW1 were used in our experiments.

We have organized the paper as follows. Implementation details along with description of datasets, Naïve Bayes classifier algorithm, and pseudocode for the proposed framework are described in Sect. 2. Section 3 has the inference along with the results of the study. Conclusions are presented in Sect. 4.

2 Implementation

2.1 Datasets from NASA MDP

We have used seven datasets from NASA Metrics Data Program (MDP) for our experiments. NASA had kept the metrics data along with problem and product data of MDP as open database for the public consumption. Many researchers have leveraged the datasets from MDP for conducting the research on predicting software defects [1, 5–12].

MDP repository has metrics of their software along with the error data indicating whether the module is defective. The error data is represented by a binary attribute. The attribute has a value of "yes" if the corresponding module has bugs else the attribute holds "no" as its value. Halstead, McCabe, LOC, error, and requirement metrics are included in the repository.

Table 1 depicts the characteristics of the seven datasets used for our research. KC3 has highest number of attributes 40 followed by PC3 and MW1 with 38 attributes. PC2 has 37 attributes. PC1, CM1, and KC1 contain 22 which is the lowest among the datasets used.

All datasets have lines of code above 20,000 except MW1 which has 8000 lines of code. KC1 has the highest LOC of 43,000. Highest number of modules was 4505 for PC2 followed by KC1 with 2107. The lowest was for MW1 with 403.

Table 1 Dataset characteristics

Datasets	Number of attributes	Lines of code	Programming language	Number of modules	Modules with low risk	Modules with high risk
PC1	22	40,000	C	1107	1031	76
PC2	37	26,000	C	4505	4482	23
PC3	38	40,000	C	1563	1403	160
CM1	22	20,000	C	505	457	48
KC1	22	43,000	C++	2107	1782	325
KC3	40	25,000	Java	458	415	43
MW1	38	8000	C	403	372	31

2.2 Naïve Bayes Classifier

Naïve Bayes classifier is based on Bayes theorem which works on conditional probability. Bayes theorem finds the probability of an event occurring based on the probability of another event that has already occurred. Mathematically, it is represented by the following equation

$$P(y|X) = \frac{P(X|y) * P(y)}{P(X)} \tag{1}$$

y is a class variable, and X is a dependent feature vector (of size n) where

$$X = x_1, x_2, x_3, \ldots x_n \tag{2}$$

The posterior probability of $P(y|X)$ is calculated from the prior probability $P(y)$ with $P(X)$ and $P(X|y)$.

After calculating the posterior probability for many different hypotheses, the one with the highest probability is selected. The maximum probable hypothesis is called the maximum a posteriori (MAP) hypothesis which can be written as

$$MAP(y) = \max \frac{P(X|y) * P(y)}{P(X)} \tag{3}$$

which is used to make predictions for the new data in Naïve Bayes model.

2.3 Proposed Framework

We have used Naïve Bayes classifier for conducting the experiments using the datasets described in Table 1. It has been reported by researchers [11] that Naïve Bayes classifier performs better than other learners with the NASA defect dataset.

As the datasets contain many variables, applying learning algorithms to the complete dataset returns poor accuracy. We need a mathematical procedure to find few important variables from a dataset with a motive to capture as much information as possible. Principal component analysis (PCA) [13] helps to overcome such difficulties.

PCA is a powerful statistical tool which is used to extract important variables from a large set of variables contained in a dataset. PCA transforms several correlated variables into a smaller number of uncorrelated variables called principal components. The principal components of a dataset are the eigenvectors of a covariance matrix and are orthogonal. These eigenvectors are associated with an eigenvalue which is interpreted as the magnitude of the corresponding eigenvector.

An eigenvector is a direction of the line, whereas eigenvalue is a number which denotes how the data is spread out on the line. The eigenvector with the highest eigenvalue is the principal component.

It is a general practice to split the data into three parts and use one part for testing and two parts for training. To make the sample that is used for testing representative, each class in the full dataset is represented in right proportion in the training and testing sets. The procedure called stratification takes care of representing the class in both training and testing sets through random sampling.

Hold-out is a method using which a little bit of dataset is held out for testing and the rest is used for training. Using the hold-out method with different random seeds, each time is called "repeated hold-out." Cross-validation is a way of improving upon repeated hold-out. It is a systemic way of doing repeated hold-out that improves upon it by reducing the variance of the estimate. Figure 1 describes the pseudocode for tenfold cross-validation.

Step 1:	Load the data
Step 2:	Randomize the data
	2.1 Create a Random seed generator
	2.2 create a copy of the original data
	2.3 randomize data with number generator
Step 3:	Generate the folds
	3.1 Divide the data into folds
	3.2 for each k = 1, 2, 3,…10, fit the model with the parameter λ to the other 9 folds giving $\hat{\beta}^{-k}(\lambda)$ and compute its error in predicting the kth part:

$$E_k(\lambda) = \sum_{i \,\varepsilon\, kth\, part} (y_i - x_i\, \hat{\beta}^{-k}(\lambda))^2$$

Step 4:	Compute the average error and choose the value of λ that makes $CV(\lambda)$ smallest

$$CV(\lambda) = \frac{1}{K}\sum_{k=1}^{K} E_k(\lambda)$$

Fig. 1 Pseudocode for 10-fold cross-validation

```
Input:      Datasets from NASA Metrics Data Program: Datasets = {PC1, PC2, PC3, CM1, KC1, KC3, MW1}
Input:      Naïve Bayes Classification Algorithm:  Learning_Alg = Naïve Bayes
Dimensionality Reduction:       Principal Component Analysis (PCA)

For each dataset in Datasets do
            full_ds  = Datasets
            Apply Stratified 10 fold Cross Validation for full_ds
            Calculate Prediction Accuracy for Learning_Alg using full_ds
            reduced_ds = PCA(Datasets)    // get Dimensionality reduction using PCA
            Apply Stratified 10 fold Cross Validation for reduced_ds
            Calculate Prediction Accuracy for Learning_Alg using reduced_ds
EndFor

Output:
a.          Reduced datasets using PCA for 7 Datasets
b.          Accuracy of Naïve Bayes classifier for 7 Datasets
```

Fig. 2 Pseudocode for the proposed framework

The prediction model using Naïve Bayes is applied to the full datasets as well as to the reduced datasets generated using PCA. The performance of Naïve Bayes is in both scenarios compared.

Figure 2 details the pseudocode for the proposed framework. Experiments were conducted using WEKA data mining toolkit [13].

3 Results and Discussion

Publicly available datasets PC1, PC2, PC3, CM1, KC1, KC3, and MW1 from NASA Metrics Data Program are used for our experiments. Naïve Bayes classification algorithms are used for prediction modeling. PCA was used for selecting the features from the datasets.

Results for dimensionality reduction through PCA are depicted in Table 2. Minimum reduction was 59% for PC1 followed by 63% for PC3. Sixty-four percentage reduction in features was achieved for CM1 and KC1. MW1 had 68%, PC2 had 70%, and KC3 had the highest with 73%.

Experiments were conducted for seven datasets using Naïve Bayes as learning algorithm with full feature set and then with reduced feature set.

Figure 3 details the comparison of performance using Naïve Bayes algorithm with and without PCA. We have witnessed tremendous increase of 155% in prediction accuracy when the reduced feature set of PC3 was used rather than the full dataset. MW1 dataset had an increase of 3.25% in prediction accuracy followed by PC2 and CM1 with 1.65% increase. KC3 had 1% increase; KC1 and PC1 had 0.6% increase in prediction accuracy over full dataset.

Table 3, column 2, depicts the results of experiments conducted with full datasets. Naïve Bayes had top performance for PC2 with 95.5% prediction accuracy followed by PC1 with 89% accuracy. Eighty-five percentage accuracy was achieved for CM1

Table 2 Dimensionality reduction for datasets using PCA

Datasets	Number of features in the original dataset	Number of features in the reduced dataset	% Reduction
PC1	22	9	59.09
PC2	37	11	70.27
PC3	38	14	63.16
CM1	22	8	63.64
KC1	22	8	63.64
KC3	40	11	72.50
MW1	38	12	68.42

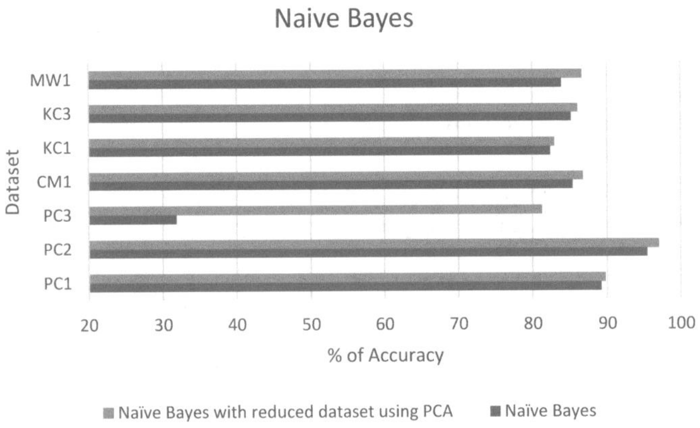

Fig. 3 Performance of Naïve Bayes

and KC3. MW1 had an accuracy of 83.9% and KC1 with 82.4%. PC3 had the lowest prediction accuracy of 31.8%. Overall average accuracy for the full datasets using Naïve Bayes classifier was 79%.

In Table 3, column 3, we have detailed the performance of Naïve Bayes classifier with reduced datasets. PC2 had the highest prediction accuracy of 97% followed by PC1 with 89.7%. Prediction accuracy of 86.7% was achieved for CM1 dataset. Accuracy of 86.6% for MW1, 86% for KC3, and 82.9% for KC1 has been noted. PC3 had the lowest prediction accuracy of 81.2% but performed very well in comparison with full dataset usage. It is heartening to note that the prediction accuracy of all datasets was greater than 80% when the reduced feature sets were used along with Naïve Bayes classification algorithm. Overall average accuracy achieved was 87.2% which is 10.3% higher than the accuracy achieved with full datasets.

Table 3 Performance of Naïve Bayes algorithm with full and reduced datasets

Datasets	Naïve Bayes with full dataset (% of accuracy)	Naïve Bayes with reduced dataset using PCA (% of accuracy)
PC1	89.1794	89.7205
PC2	95.4574	97.0347
PC3	31.8222	81.2444
CM1	85.3414	86.747
KC1	82.3613	82.8829
KC3	85.1528	86.0262
MW1	83.871	86.6005
Average accuracy	79.0265	87.1795
% of increase		10.31673824

4   Conclusion

Research on software defect prediction is gaining more importance day by day as the insights into the prediction will help the software engineering teams to "do more with less" [1] and thereby improve their productivity and quality of the software.

Dimensionality reduction using PCA along with Naïve Bayes algorithm has been proposed for predictive modeling. It is recommended to use stratified tenfold cross-validation while building the model.

Experiments were conducted using seven datasets PC1, PC2, PC3, CM1, KC1, KC3, and MW1 from NASA Metrics Data Program. An average increase of 10.3% prediction accuracy was witnessed when the classifier is used with datasets containing selected features from PCA.

Future research would include finding ways to eliminate class imbalance problems and exploring combination of machine learning techniques with various feature extraction methods. We also plan to use datasets from multiple domains.

References

1. Dhamayanthi N, Lavanya B (2019) Improvement in software defect prediction outcome using principal component analysis and ensemble machine learning algorithms. In: Hemanth J, Fernando X, Lafata P, Baig Z (eds) International Conference on Intelligent Data Communication Technologies and Internet of Things (ICICI) 2018. ICICI 2018. Lecture notes on data engineering and communications technologies, vol 26. Springer, Cham. https://link.springer.com/chapter/10.1007/978-3-030-03146-6_44
2. Murillo-Morera J, Castro-Herrera C, Arroyo J, Fucntcs-Fernandez R (2016) An automated defect prediction framework using genetic algorithms: a validation of empirical studies. Inteligencia Artif 19(57):114–137

3. Song Q, Jia Z, Shepperd M, Ying S, Liu J (2011) A general software defect-proneness prediction framework. IEEE Trans Softw Eng 37(3):356–370
4. Shirabad JS, Menzies TJ (2005) The PROMISE repository of software engineering databases. School of Information Technology and Engineering, University of Ottawa, Canada. Available: http://promise.site.uottawa.ca/SERepository
5. Wang S, Ping HE, Zelin L (2016) An enhanced software defect prediction model with multiple metrics and learners. Int J Ind Syst Eng 22(3):358–371
6. Shatnawi R, Li W (2016) An empirical investigation of predicting fault count, fix cost and effort using software metrics. (IJACSA) Int J Adv Comput Sci Appl 7(2)
7. Menzies T, Greenwald J, Frank A (2007) Data mining static code attributes to learn defect predictors. IEEE Trans Softw Eng 33(1)
8. Jiang Y, Lin J, Cukic B, Menzies T (2009) Variance analysis in software fault prediction models. In: 20th international symposium of software reliability engineering
9. Koru AG, Liu H (2005) An investigation of the effect of module size on defect prediction using static measures. In: Promise '05
10. Singh P, Verma S (2014) An efficient software fault prediction model using cluster based classification. Int J Appl Inf Syst (IJAIS) 7(3)
11. Zhang H, Nelson A, Menzies T (2010) On the value of learning from defect dense components for software defect prediction. In: Promise 2010, 12–13 Sept
12. Jin C, Dong E-M, Qin L-N (2010) Software fault prediction model based on adaptive dynamical and median particle swarm optimization. In: Second international conference on multimedia and information technology
13. Witten IH, Frank E (2005) Data mining, practical machine learning tools and techniques. Morgan Kaufmann, San Francisco

Person Tracking and Counting System Using Motion Vector Analysis for Crowd Steering

K. Sujatha, S. V. S. V. P. Rama Raju, P. V. Nageswara Rao, A. Arjuna Rao and K. Shyamanth

Abstract Video surveillance has been in use since a protracted time as an assistance to beat security and other problems. Historically, the video outputs area unit monitored by human operators and area unit sometimes saved to tapes for later use. Sensitive areas like shopping malls, banks, huddled public places want a strict police investigation and may require management of the flow of individuals mechanically. To do such automation, a wise video closed-circuit television is required for today's world equipped with machine learning algorithms. In this project, a sensible visual closed-circuit television with person detection and following capabilities is bestowed. This can be used to regulate the flow of persons into the sensitive areas, which is often achieved by count the persons who are getting into and going through these areas, so knowing the overall capability a sensitive space is holing at any specific purpose of your time. Motion vector analysis is that the main construct that is used here to realize the following of the persons. This has a tendency to count the persons who are getting into and going out stationary cameras fixed points, the capability is obtained as distinction between the count of the persons entered and count of the persons who left the sensitive space. Any sensitive space would have a restricted house to accommodate. So it is necessary to prohibit the persons from getting into sensitive space, once the capability is reached to threshold price.

K. Sujatha (✉) · S. V. S. V. P. Rama Raju
CSE Department, Dadi Institute of Engineering and Technology, Visakhapatnam, India
e-mail: sujathakota29@gmail.com

S. V. S. V. P. Rama Raju
e-mail: ramarajusvsvp@gmail.com

P. V. Nageswara Rao
CSE Department, GITAM University, Visakhapatnam, India

A. Arjuna Rao
Miracle Educational Society Group of Institutions, Visakhapatnam, India

K. Shyamanth
Andhra University, Visakhapatnam, India

© Springer Nature Singapore Pte Ltd. 2019
N. Chaki et al. (eds.), *Proceedings of International Conference on Computational Intelligence and Data Engineering*, Lecture Notes on Data Engineering and Communications Technologies 28, https://doi.org/10.1007/978-981-13-6459-4_25

249

Keywords Video surveillance · Sensitive areas · Motion vector analysis · Moving person detection and tracking · Background subtraction · Person counting

1 Introduction

Efficient and reliable automatic person following and count are often terribly helpful for many industrial applications used for security and other people management. This may be used to guarantee safety of the folks within the sensitive areas. Areas like public places, shopping malls, and banks are often secured from being overcrowded. To monitor these places, this needs the variety of procedures involving cameras in standard police investigation systems. This also needs additional human operators and therefore the storage devices with high volumes of data and created. It is impracticable to confirm correct watching of sensitive areas for long times.

Large numbers of individuals are on the road in sensitive areas, and rate of those folks inside a fixed region might indicate the number of persons crossing this region inside a precise amount of time. It is arduous to count the persons after they enter or leave during a cluster at a precise place. This downside is often solved by the employment of quick growing PC vision technology and high-speed PC as a processor for counting these persons. Persons are often counted by taking continuous footage by video camera and spending this real-time footage unceasingly to PC through USB connective for image process. This image process composes of segmentation, background elimination, and blobs detection, etc., and may be done by using sophisticated image processing techniques; background subtraction techniques largely observe motion in several real-time vision police work applications.

These approaches think about variations between the incoming and background pictures to observe foreground objects. Background subtraction provides the foremost complete feature knowledge; however, it is very sensitive to dynamic scene alterations because of illumination changes and extraneous events. Researchers are currently dedicated to developing a study background model to forestall falseness in motion detection caused by scene changes. The background model is sporadically updated employing a combination of pixel and object-based strategies. Background subtraction speed and accuracy rely upon the results of the background extraction algorithmic program. If there is a scene (frame) with none moving object within the image sequences, it is often a background frame with a pure background image. However, within the globe, there is hardly to induce a pure background image. In such systems, as in several alternative applications, extracting a pure background image from videos that embody moving objects is important [1, 2].

Conventional technique of person enumeration system is manual enumeration. The complexity lies in counting the quantity of persons passing a space by victimization counter. Even to count few persons it is not sure that will be counted accurately at intervals in a brief amount of time. Manual enumeration is labor intensive and extremely pricey. Human labors have restricted span and dependableness once great amount of knowledge has got to be analyzed over a protracted amount of time, particularly in jam-panicked conditions. It is conjointly arduous to deliver physical leads to real time for online police investigation. Consequently, it is necessary to develop the automatcd person enumeration system and this is often not an easy task, there are some things troublesome to unravel even with today's technology [3].

Person enumeration could be a difficult scientific downside and associated with several sensible applications like railway platform, enumeration of persons within the elevator, trains, enumeration of the quantity of persons passing through security doors in searching malls, enumeration of variety of persons presents in chain store, enumeration of variety of persons visit the recreation park and enumeration the quantity of persons operating within the laboratory. One in every one of the automatic ways of enumeration persons is that the small controller-based enumeration system. Small controller-based system unremarkably used for enumeration for little scale, except for massive scale and business use, and this method has some limitations like lack of accuracy. Conjointly, unbearable sensors will be wont to count persons, unbearable receivers will count the quantity of the persons once it detects the echo bouncing removed from the persons at intervals the detection zone [4].

The accuracy of enumeration degraded once several objects walk across the detection region, particularly person before of the sensors blocked the detection of different persons. Conjointly, microwave sensors and weight-sensitive sensors are one in every one of the devices which will be used to count persons. Owing to fast development of laptop and laptop vision system, it is doable to count persons victimization computer vision even if the method is extraordinarily pricey in terms of computing operations and resources. In general, enumeration persons are vital in police investigation-based mostly applications, marketing research, and persons management. Persons detection by suggests that of artificial vision algorithms which is an energetic analysis field. A detailed analysis of the method conferred within the paper can show that by this method persons will be counted effectively [5].

2 Related Work

Due to its generality in varied contexts, background subtraction has been the main target of a lot of analysis. Some use a background subtraction model engineered from order statistics of background values throughout a coaching amount to implement an applied math color background rule that uses color chrominance and brightness distortion. The various background subtraction ways are often classified following the model utilized in the background illustration step.

Basic background modeling is done by the common, median or bar chart analysis over time. Statistical background modeling: the only Gaussian, Mixture of Gaussians or Kernel Density Estimation. Background estimation is done by Wiener, Kalman or hebychev filters [6].

R. Jain and H. Nagles, 1979, given the earliest approach to background sub-traction within the late 1970s, mistreatment frame differencing to notice moving objects. Ulterior approaches projected a progression of probabilistic models to han-dle uncertainty in background look. Researchers have recently given some classical algorithms for background extraction, as well as the mean, median, stable interval determination, amendment detection, and mode algorithms. Researchers have addi-tionally used some extremely advanced ways in background extraction algorithms, including textural and statistical features and genetic algorithms

Researchers have proposed edge-based foreground segmentations that use color information or both color and gradient information. The Wallflower algorithm aims to solve many problems in background maintenance, including varying lighting con-ditions, by developing a three-level system: pixel-, region- and frame-level. More recently, researchers have investigated the scale-invariant feature transform (SIFT), as it can produce tracking features that are stable across frames with a certain amount of camera/object viewpoint variation. This becomes important when considering some combination of many features in the recognition and tracking technique and deciding on the best features to use under different circumstances is necessary [3, 6].

Whether region-based methods are better than edge-based method remains unknown. However, it is generally acknowledged that the edge-based methods often provide better boundary localization. A flow field is developed based on magneto-static interactions between the active contours and object boundaries. Canny showed that the first derivative of a Gaussian can approximate the optimal detector. By con-volving the image with this filter, the edge detection is equivalent to finding the gradient magnitude maxima of a Gaussian-smoothed image in the appropriate direc-tion.

3 Person Counting System

The architecture followed by the system is shown Fig. 1 of automatic person enumer-ation system. The photographs are taken from the video camera that is assumed to be at a stationary location, the image is felt a preprocessing operation before applying motion vector analysis thereon. Once performing arts motion vector analysis, the data is felt a choice creating operation for pursuit and enumeration the persons.

3.1 Description

In detail, the method of the person investigating system is shown within the following diagram, acquisition of the image is that the initial demand at the system, that is finished by the video camera so it's remodeled into grayscale image for analyzing and resizes to a smaller resolution.

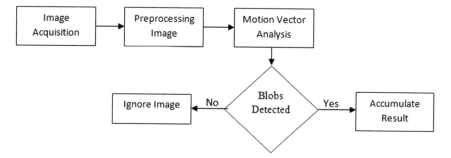

Fig. 1 Architecture of person counting system

Then it's skilled filtered method for noise removal so the noise reduced image is regenerate to binary image that is in terms of 0's and 1's, for this image the vector motion analysis is performed supported the previous image and this is often done until the last image of the video has reached. And an oblong mask is formed to point the persons within the image and blob analysis is performed to the image to spot the person and this is often mapped to motion vector and also the detected persons count in every image is keep and accumulated to a dynamic variable within the program and it is written within the image at the left corner when accumulation.

4 Results

In the following figures, the persons are viewed within the image at the side of the quantity at the left corner of the image indicating the accumulated count of United Nations agency. The overall persons who have detected within the video, and therefore, the rectangular frames bounding the persons are accustomed to indicate that the detected object could be a person within the image. Figure 2 shows the context of an image with no persons (Fig. 3)

Figure 4 depicts a context with an image with single person followed by Fig. 5 showing the count of persons crossed the specified boundary.

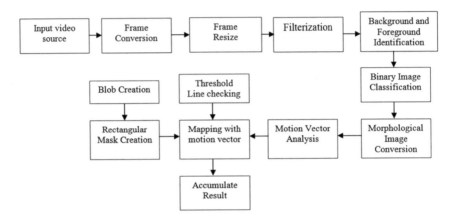

Fig. 2 Person counting system working principle

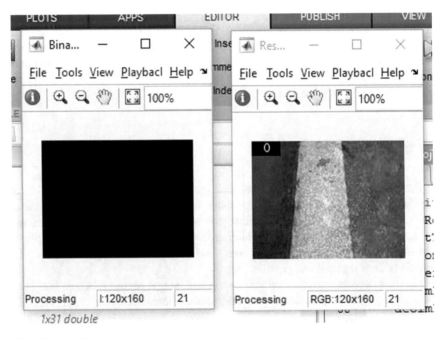

Fig. 3 Image with no persons

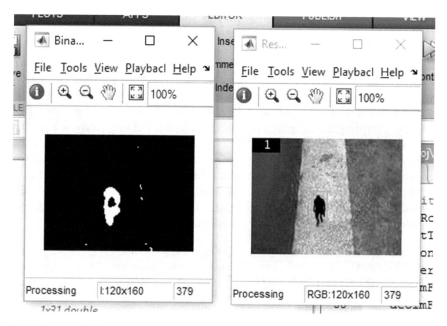

Fig. 4 Image with singe person crossed and the indicated count

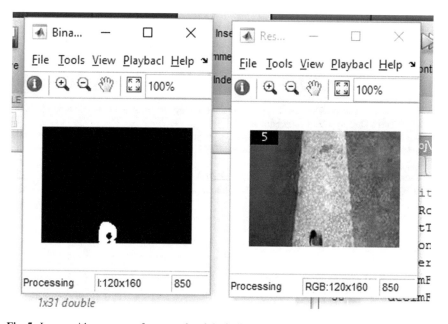

Fig. 5 Image with persons so far crossed and the indicated count of the persons

5 Conclusion

The person count system may be a combination of the background subtraction method, chase objects, and motion vector analysis. This paper shows an approach to count persons getting into through a selected place by employing a video recorded from a video camera. This person count system may be simply enforced in MAT-LAB programming with a decent configuration laptop and may ready to count the persons in real time. This method offers a good rate of potency compare thereto of the bequest device primarily based person count system. This person count system may be utilized in many sensitive places like banks, jewelry mall, supermarket, recreational park, railway platforms so several places for count persons and realize the individual's gift in a very specific place. This approach is incredibly imported for marketing research to investigate the effectiveness of business policy.

Acknowledgements The authors express deep sense of gratitude to the management of DIET College and GITAM University for the entire support and facilities provided to us throughout in bringing out this successful work.

References

1. Geronimo D, Lopez AM, Sappa AD, Graf T (2010) Survey of pedestrian detection for advanced driver assistance systems. IEEE Trans Patt Anal Mach Intell (PAMI) 32:1239–1258
2. Son B, Shin S, Kim J, Her Y (2007) Implementation of the real time people counting system using wireless sensor networks. Int J Multimedia Ubiquitous Eng 2(3):63–79
3. Wren C, Azarbayejani A (1997) Pfinder: real-time tracking of the human body. IEEE PAMI 19(7):780–785
4. Favalli L, Mecocci A, Moshetti F (2000) Object tracking for retrieval applications in MPEG-2. IEEE Trans Circ Syst Video Technol 10(3)
5. Achanta R, Kankanhalli M, Mulhem P (2002) Compressed domain object tracking for automatic indexing of objects in MPEG home video. In: IEEE international conference on multimedia and expo, vol 2, pp 61–64
6. Terada K, Yoshida D, Oe S, Yamaguchi J (1999) A method of counting the passing people by using the stereo images. In: International conference on image processing, ISBN 0-7803-5467-2

Author Index

Printed in the United States
By Bookmasters